AZO FUNCTIONAL POLYMERS

AZO FUNCTIONAL POLYMERS

Functional Group Approach in Macromolecular Design

G. Sudesh Kumar

ICI INDIA LTD.
HYDERABAD, INDIA

LANCASTER · BASEL

Azo Functional Polymers
a **TECHNOMIC**® publication

Published in the Western Hemisphere by
Technomic Publishing Company, Inc.
851 New Holland Avenue
Box 3535
Lancaster, Pennsylvania 17604 U.S.A.

Distributed in the Rest of the World by
Technomic Publishing AG

Printed in the United States of America
10 9 8 7 6 5 4 3 2 1

Main entry under title:
Azo Functional Polymers: Functional Group Approach in Macromolecular Design

A Technomic Publishing Company book
Bibliography: p.

Library of Congress Card No. 92-64256
ISBN No. 0-87762-936-6

With love to Padmaja

TABLE OF CONTENTS

ACKNOWLEDGEMENTS

I would like to take this opportunity to thank Professor M. Saffran, Medical College of Ohio at Toledo for his assistance and collaboration in azo polymer work. I am especially thankful to Dr. K. Srinivasan, Alchemie Research Centre, Thane who patiently and graciously gave his time to read various chapters of the text. I am also grateful to Mr. A. Ghogare, Mr. S. R. Gaonkar, Mr. H. Mahalingam and Mr. N. V. Joshi who helped me in my research work and literature search during my long association with azo polymers. I express sincere thanks to Mr. V. V. Joshi for drawing work and Mr. V. Y. Bal for help in preparing the manuscript.

1

FUNCTIONAL GROUP REACTIVITY IN POLYMER SCIENCE

1.1. Introduction

Polymer science as a coherent subject is barely sixty years old. It deals with polymers in terms of their chemistry, their molecular structure and physical properties, their applications, and processing into useful forms. The chemical and physical interactions of the atoms of a polymer are governed by the same laws that describe systems of small molecules, but extreme molecular size introduces a new realm of properties. The chemical basis for the special properties of polymers that equips them for so many applications and functions, both in nature and in the materials world, is therefore not to be sought in the peculiarities of the chemical bonding but rather in their macromolecular constitution, specifically in the attributes of the long molecular chains. The structural basis for the vast range of properties and behavior of macromolecules, both natural and synthetic, is the chemical composition of main chains and the type of substituents linked to them.

The diversity of polymers represented by a given chemical composition varies enormously with the number and nature of repeating units present. The chemical constitution of a polymer requires evaluation at both the macro and micro levels. At the macro level, we consider the overall architecture as expressed in the molecular weights or chain length, the molecular weight distribution, the degree of branching/cross-linking and so forth. At micro level, it is necessary to consider the chemical constitution of the building blocks of macromolecular chains and the functional groups connecting or constituting the repeating units. In combination, variation at these two levels admits a potential for an almost endless diversity of properties. The art of polymer synthesis, therefore, resides in the optimum and coordinated utilization of each element of the complete ensemble of strategy, and in the selection of appropriate macromolecular framework and polymerization process for the desired application.

Polymer constitution is always discussed in terms of polymerization mechanisms, as addition- and condensation-type polymers; and in terms of sequence and distribution of repeating units as block and graft copolymers, etc. It is also discussed on the basis of polymer chain structure, as linear, branched, and cross-linked chains, but seldom in terms of the functional groups constituting these polymers. When the polymers are classified on the basis of functional groups as polyesters, polyamides, polyurethanes, etc., most often the functional groups serve as *structural* linkages rather than *functional* linkages. To illustrate the point, most of the polyesters are seldom designed to manifest the chemical properties of ester functionality but instead for their bulk or mechanical properties such as glass transition temperature (T_g) or crystallinity. If the polyester is designed as an absorbable suture, where it undergoes enzymatic hydrolysis, the ester group can be considered as a functional linkage and the polymer as a functional polymer. As the role of polymers is changing from structural materials to functional materials, considerable effort is underway to tailor-make polymers towards specific end-uses.

Many functional groups have been incorporated into the macromolecular framework in order to design materials to perform specific chemical functions. But very few functional groups in organic chemistry exhibit a range of thermal, chemical, and biological properties, such that each of these properties has been the basis for a distinct class of functional polymers. The azo linkage is one such functional group. The present work is conceived as a novel method of illustrating the functional group approach in macromolecular design based on azo functionality. It is hoped that this volume, which illustrates the functional group approach to polymer design, the tools available to it, its potential and predictive capacity, will provide insight into the basic philosophy and strategies employed in the design and execution of polymer synthesis. I begin the polymer chemist's expedition with a historical look at the role of the functional groups in the evolution of polymer science.

1.2. Functional Groups in Polymer Science: Flory's Principle

The realization of the importance of the nature and reactivity of functional groups in polymer chemistry can be traced to the very beginnings of polymer science. When, after graduating from OSU in 1934, Flory was employed by the DuPont company as a research chemist in the famous research team of Wallace Carothers, he faced a number of interesting and fascinating problems relating to polymer synthesis and behavior.[1] One of them dealt with the formation of condensation polymers by successive chemical reactions between carboxyl and hydroxyl groups to form a linear polyester. In order to follow such processes quantitatively and to know the molecular weight distribution during the reaction, it was necessary to assess correctly the intrinsic reactivity of the functional groups. It was commonly supposed that the normal reactivity of a given kind of functional group would be suppressed if it were on a very large molecule; mere size *per se* was considered to impart sluggishness that would bar unlimited chain growth, and existing polycondensation rates were so interpreted. Flory, however, in constructing a straightforward statistical treatment of the distribution problem, took the contrary view that the reactivity, under given conditions of solvent, temperature, pressure, and concentration is essentially a function of local structure and not of overall molecular size.[2,3]

The physical and chemical properties of macromolecular systems are not fundamentally *sui generis*, but are understandable to the same degree as those of nonpolymeric matter in light of the basic disciplines of thermodynamics, kinetics, etc. The polymerization kinetic analysis is greatly simplified if one assumes that the reactivity of one functional group of a bifunctional reactant is the same irrespective of whether or not the other functional group has reacted, and the reactivity of a functional group is independent of the size of the molecule to which it is attached. These simplifying assumptions are justified on the basis that many step growth polymerizations have reaction rate constants that are independent of the reaction time or polymer molecular weight. The independence of the reactivity of a functional group on molecular size can be observed from the reaction rates in a homologous series of compounds differing from each other only in molecular weight. It is evident that although there is a decrease in the reactivity with increased molecular size, the effect is only significant at a very small size. The reaction rate constant very quickly (at $n = 3$) reaches a limiting value that remains constant and independent of molecular size.

There is ample theoretical justification for the proposition that the reactivity of a functional group is independent of molecular size. The fairly common misconception regarding the alleged reactivity of groups attached to large molecules comes from the lower diffusion rates of the latter. However, the observed reactivity of a functional group is dependent on the collision frequency of the functional group, and not on the diffusion rate of the whole molecule. The collision frequency is the number of collisions one functional group makes with other functional groups per unit time. A terminal functional group attached to a growing polymer has a much greater mobility than would be expected from the mobility of the polymer molecule as a whole. In addition, this functional group has appreciable mobility due to the rearrangements that occur in nearby segments of the polymer chain. The collision rate of such a functional group with neighboring groups will be about the same as for small molecules.

The net result of these considerations is that the reactivity of a functional group will be independent of the size of the molecule to which it is attached. Exceptions to this occur when the reactivities of the groups and/or molecular weights of the polymers are very high. The polymerization becomes diffusion-controlled in these cases because mobility is too low to allow maintenance of the equilibrium concentration of reactive pairs and of their collision frequencies. It is important to note that "the principle of equal reactivity of the functional groups" as a basis for polymerization kinetic treatment has coincided with the birth of polymer science and has evolved with the growth of contemporary polymer science.

1.3. Polymer-Anchored Functional Groups: Merrifield Synthesis

Until the mid-1960s, synthetic polymers were of considerable technological interest largely as materials rather than as organic molecules in their own right. The synthesis of new polymers and the detailed investigation of the mechanisms of existing polymerizations were mainly directed towards the production of improved materials. At the same time, techniques for the characterization of macromolecules had been translated into commercially available instrumentation, leading to a rapid expansion in the material science study of polymers and the search for a molecular understanding of polymer physics. Since the mid-1960s, synthetic macromolecules have been increasingly recognized as organic species, capable of behaving as organic reactants and susceptible, under appropriate conditions, to all the chemical tranformations of smaller organic species. The rediscovery of polymers as organic molecules and their use in organic synthesis was made by Merrifield in 1963[4] when he introduced solid phase synthesis. Since then, the second wave of functional group-polymer interrelationships has dealt with the chemical transformations of polymer-bound functional groups.[5-8] Polymers were prepared that contained almost every functional group found in low molecular weight compounds.

When a reagent (or a substrate) is covalently bound to a polymer, it acquires the physical properties of the latter. Consequently, the functionalized polymer (if reasonably cross-linked) remains insoluble in common organic solvents. If the polymer is porous and swells in a suitable solvent, the functional groups anchored on it are easily available for chemical transformations. Covalent attachment of the functional groups to a polymer helps in keeping track of the transformed product in a chemical synthesis, resulting in simplification of the processes. A classification of polymer-mediated functional group reactions basically falls into two categories.

(1) The first type includes reactions in which the polymer acts as a carrier (support) for the substrate (functional group). The product remains attached to the support while the by-products, excess reagents, and solvents all remain in solution and can be removed by filtration. The synthesis may involve a single step (such as acylation of an enolizable polymeric ester) or it may be a sequential synthesis of a biopolymer, as in peptide synthesis where the successive addition of monomers is carried out as a graft on the basic polymer chain. The last step in such a synthesis involves cleavage of the product from the polymer backbone.

(2) The second type includes reactions in which a functionalized polymer supporting a conventional synthetic reagent is reacted with a low molecular weight substrate that is transformed into the product. The excess of polymeric reagent and the spent polymer remain insoluble, whereas the product goes into the solution.

The number of publications in this area since the early 1960s is too large even for a brief coverage. A number of excellent books, reviews and monographs[5-14] have dealt with the subject. In general, polymers undergo chemical reactions in much the same way as do their low molecular weight analogs, provided the site of reaction is accessible. For example, carboxylic polymers readily undergo esterification, amidification, peracid formation, etc. The carbonyl group in polymers undergoes its usual reactions like oxidation and reduction. Many of these reactions have been used for preparing functionalized polystyrene-based reagents. However, it is increasingly recognized that, when polymers are used as supports for catalysts or organic reagents, the reactivity and selectivity of the supported catalysts or reagents may be seriously changed by the so-called "polymer effects," the origin of which may be physical (viscous diffusion effects, steric effects, local concentration effects) or chemical (microenvironmental interactions). Based on these extensive studies, detailed structure-activity correlations have emerged governing the functional group reactivity of polymer-bound substrates. It must be said that the discussion here is only representative and not exhaustive and the reader is referred to the original reviews for detailed discussion.

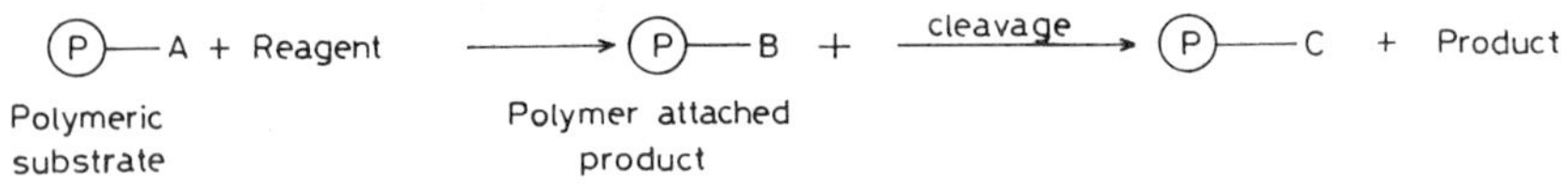

Ⓟ—A + Reagent ⟶ Ⓟ—B + $\xrightarrow{\text{cleavage}}$ Ⓟ—C + Product

Polymeric substrate — Polymer attached product

FIGURE 1.1

1.4. General Features of Polymer-Anchored Functional Group Reactions

The study of the chemical reactions of polymers has emerged as one of the most active fields in polymer science because of its unique ability to produce speciality polymers with desirable chemical and physical properties through modification of readily available polymers. The field is extremely diverse and has grown to encompass many topics.[10-12] The use of functional group transformations to introduce required functional or reactive groups, to alter polymer surfaces, to provide grafts and unusual side chain substituents, and to aid in the analytical characterization of polymers, represents a significant portion of the scope of this scientifically interesting and technologically useful field of research.

To consider a polymer chain as a sequence of pendant functional groups strung from an extended backbone chain is an oversimplification. Likewise, organic reactions are normally written and often discussed as if they occurred between isolated molecules of the reactants. Although these representations are ideal for the purpose of discussion, the influence of the microenvironment, i.e., the macromolecular surroundings of the reacting species, is very important.

The most important factors in polymer-anchored functional group transformations, which are relevant from the point of view of designing novel functional polymers, are:

- the nature of the polymeric support
- location of the functional group
- steric effects
- polymer morphology
- microenvironmental effects
- conformation of the polymer chain

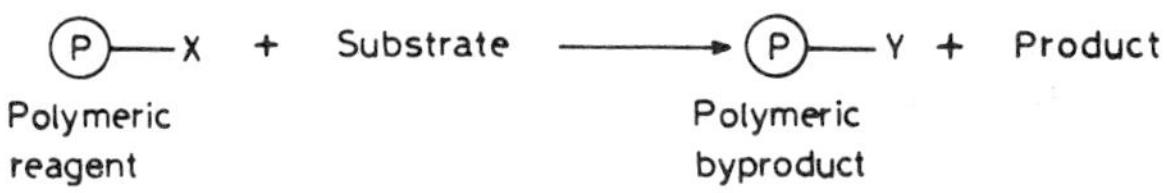

Ⓟ—X + Substrate ⟶ Ⓟ—Y + Product

Polymeric reagent — Polymeric byproduct

FIGURE 1.2

Many of these factors influence the polymer properties in different ways, depending upon the functional group. These effects, as relevant to azo functionality, would be evident in the subsequent chapters. As a result, I choose not to get into the details of polymer design, but a few general observations are made here with regard to the design criteria of functional group-containing polymers.

1.4.1. The Nature of the Polymeric Support

The variety of functionalized polymers has grown both in the nature of the polymer backbone and in the array of functional groups.[15] Functionalized polystyrene has been the support most widely used, first as an ion-exchange resin, and later as a polymer-bound reagent or catalyst in a variety of chemical reactions, e.g., acylation, alkylation, and oxidation-reduction. Polystyrene is well suited to chemical modification since the aromatic rings can be easily functionalized, and the polymer backbone possesses the necessary stability to allow chemical transformations without degradation. Modifications of polystyrene have included electrophilic substitution and metallation with a subsequent reaction with a nucleophile. Chloromethylated polystyrene is even more suited to chemical modification since the chlorine is easily displaced by nucleophilic reagents.[6] A variety of benzylic substituents have been demonstrated, e.g., amines, quaternary ammonium or phosphonium salts, ethers, esters, sulfonamides, alkyl sulfides, silanes, and benzophenones. Other reactions on the polystyrene backbone include diazotization,[16] and carbonylation[17] to produce blocked isocyanates. The resulting polymers have found use as reagents, chelating materials, ion-exchange resins, and photoreactive polymers.

Poly(vinyl chloroformate)[15] has been shown to undergo facile modification to provide new functional polymers, since it can react with nucleophilic compounds containing labile hydrogen atoms, e.g., alcohols, phenols, and amines. Since well-defined structures and high molecular weight polymers can be prepared, poly(vinyl chloroformate) is well suited to kinetic and mechanistic studies, in addition to its use as a substrate in preparation of reactive polymers.

Polydienes and other unsaturated polymers have also been widely used as substrates for chemical modification to produce reactive polymers. The reactions have included epoxidation, cyclization (e.g., with carbene), phosphorylation, hydroformylation, hydroxylation, halogenation, and nitration, resulting in new materials for polymer-bound reagents, functional polymers for grafting, etc.

Other addition polymers, for example, polyolefins, poly(vinyl acetate), poly(vinyl alcohol), and acrylates have also been subjects of study. Poly-(vinyl acetate) and poly(vinyl alcohol) continue to be desirable substrates for modification when highly hydrophilic materials are desired. Introduction of reactive groups for biological applications, e.g., immobilizing enzymes, has been a particularly active area of the field. Chemical modification of acrylic polymers has included hydrolysis of the ester to produce poly(acrylic acid) and poly(methacrylic acid), etc., and reaction of functional acrylic polymers, e.g., ring opening of the epoxide ring of glycidyl methacrylate-ethylene dimethacrylate copolymers to produce selective chelating materials.

Condensation polymers have received increasing attention recently as substrates for chemical modification, since they can provide superior mechanical properties compared to the addition polymers that have been so widely used. Condensation polymers that have been used historically include polyamides, polyesters, and polyurethanes. More recently, the focus has been on poly(phenylene oxides), polysulfones, and phenoxy resins. Electrophilic modification can be used on those polymers that contain activated aromatic groups, including polysulfones, phenoxy resins, and poly(phenylene oxide). The electrophilic reactions have included nitration, bromination, chloromethylation, sulfonation, amination, acylation, among others. The resulting polymers have shown utility as polymeric supports, polymeric reagents, and permselective materials.

Clearly, several factors are important in selecting the type of polymer to be used to support a given reactive group. These include the ease of preparation of potentially suitable polymers with the appropriate functional groups and the ease with which these polymers can be obtained in required physical form. In principle, the synthesis of a polymeric support should allow the control over functional group concentrations and distributions as well as cross-link density. In the case of cross-linked gels, the reactivity and the swelling characteristics can be regulated by the choice of chemical groups. Support morphology, and therefore thermal and mechanical stabilities, should be controlled by the conditions of synthesis.

It is also important that, apart from those reactive groups, the polymer should be chemically inert under the conditions of use. The polymer should provide the required microenvironment for that particular application property, such as polarity, hydrophilicity, hydrophobicity, or microviscosity.

1.4.2. Location of the Functional Group

The major factors influencing chemical reactions on polymers are the reactivity of the functional group on the polymeric matrix and its accessi-

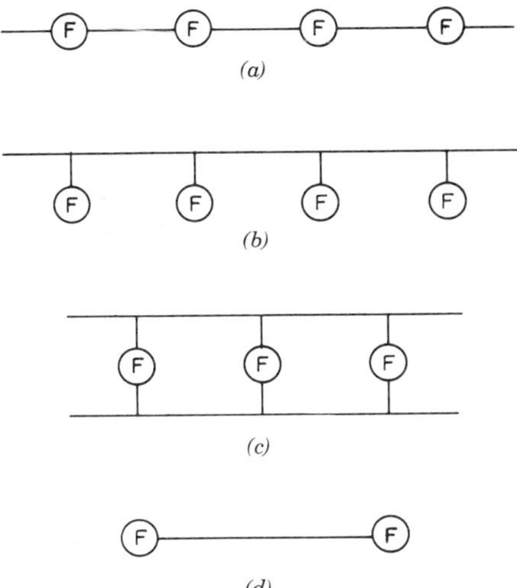

FIGURE 1.3 Functional group (F) positions in a polymer chain.

bility.[10] The functional group may be a part of the backbone, chain end, side chain, or may be present in cross-links (Figure 1.3). The synthetic methodologies available for polymer synthesis differ widely depending on the position of the group. The restrictions acting on the chain segment are different in each case. Likewise, effective contribution of the functional group to the overall macromolecular behavior is accordingly different. For example, the contribution of each functional group to polymer conformation, solubility, stability, degradation etc. from a backbone will be different from the one in side chain or crosslinks. The ability to participate in chemical reactions such as oxidation or reduction will also vary with the position of the functional group.

In general, synthetic linear molecules with backbone functionalities are to be formed by some type of polycondensation while the ones with side chains are made by addition polymerization. It is the backbone of any polymeric material which decides its essential features, whether it is aliphatic or aromatic, carbon-chain or hetero-chain, organic or inorganic, carbocyclic or heterocyclic. For example, azo functional groups have been selectively attached to side chains, in polymer backbones or in the cross-links. By site-specific photolabelling, differences in the photoisomerization rates of azobenzene units were found at the end of the polymer chain,[18] along the chain and at irregularities within the polymer matrix. These differences were rationalized by taking into account that near a crosslink junction the restrictions acting on a chain segment mainly arise from four branches while only two branches act on a segment in the chain.

1.4.3. Steric Effects: Effect of Spacer Groups

Functional groups bound to polymers might be expected to be sterically crowded, but this is probably only the case when they are very close to the backbone. As the functional groups are separated from the backbone by spacer groups, steric effects would be expected to disappear rapidly. Most polymer supported reactants are prepared from polystyrenes and here the benzene ring will act as a rigid spacer group.[17] Under conditions where the polymer backbone and the functional group interact favorably with the reaction solvent, functional groups bound to para positions of the polystyrene would not be expected to behave any different sterically than their low molecular weight analogs.

Typically, the spacers are difunctional molecules with various chain lengths. One of the chain ends is first reacted with the polymer, which of course must carry a suitable functional group. The second chain end may then be a desired functional group or it may be reacted with a small molecule carrying the desired functional group.

As shown by Regen's study[19] of the shape of the ESR signal of stable nitroxyl radicals, the mobility of the functional groups is much lower, for the same crosslinking ratio, when they are directly attached to polymer backbones than if they are dissolved in swollen particles. With this in mind accessibility and the reactivity of such groups should also be restricted. Indeed, it has often been observed that the rate of reactions involving supported reagents is enhanced when these reagents are anchored to the support through spacer groups. Moreover, spacer groups have been applied

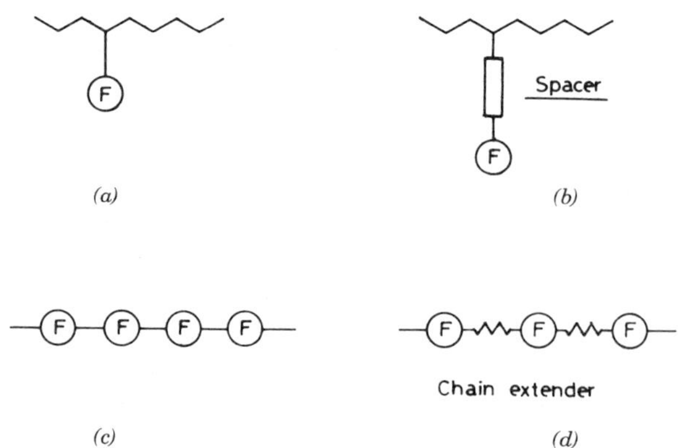

FIGURE 1.4 Polymer chains with pendant spacer units and backbone chain extender units.

more and more in the design of polymeric systems aimed at speciality applications[20] such as drug delivery systems, latex supported immunoassays and comb-like liquid crystals.

When the functional group is attached to a polymer chain reacting with a species of low molecular weight, the reaction site close to the polymer backbone will have very different solvent properties than the bulk of the solvent. We might then expect any reaction that is sensitive to medium effects to exhibit characteristic differences between the polymeric species and low molecular weight analogs. This is explained in terms of "local solvent medium."[21] It is expected that effects of the "local solvent medium" on the reactivity of polymer substituents should decay rapidly as the distance of the reaction site from the backbone is increased.

1.4.4. Polymer Morphology

When reactions are carried out under conditions such that portions of the polymer remain crystalline, only the functional groups in amorphous regions are available for reaction. The functional groups in crystalline regions will generally be inaccessible to chemical reagents and the reaction will be limited to those amorphous regions. The reaction of a polymer can be exactly analogous to that of its low molecular weight analog when it is carried out in solution by the appropriate choice of solvent and temperature. The "polymer effects" may become apparent as one moves from dilute solutions to concentrated solutions and from solution state to solid state.

Likewise, conventional polymers carrying a few percent of carboxylic, phosphoric or sulphonic acid groups when neutralized by a metal cation exhibit strong columbic interactions, which impart on them strong mechanical properties. The acid groups tend to agglomerate into hydrophilic clusters, which may be connected together by hydrophilic channels. When catalysis involves the formation of metallic particles (upon reduction of the transition metal ion) very fine particles with high catalytic activity are produced.

In most of the "supported chemistry" and in a number of functional polymers, the polymeric support/carrier is a crosslinked gel. These supports are able to swell to varying degrees in different solvent media. Partition of the reactant and products between the liquid phase and the polymer phase must then be considered. Due to the heterogeneity of the crosslink density inside the polymer phase, the accessibility of the functional group may be very different in polymer microdomains and may not necessarily be regular along the radius of the spherical bead.

1.4.5. Microenvironmental Effects

The intrinsic reactivity of a functional group does not change just because the group is attached to a macromolecule. Thus the intrinsic reactivity of for example, a polymer-supported phosphine residue will be essentially the same as that of 4-diphenylphophinylisopropylbenzene.[5] With a polymer supported reactant, however, the functional groups are usually confined to a much smaller volume than they are in the analogous low molecular weight reaction system. Interactions between the groups are potentially greater. The classical example of reactions where the presence of ionic groups alters the chemical reactivity of neighboring groups are titrations of polyacrylic acids and polymethacrylic acids with alkali and the alkaline hydrolysis of a polyacrylate, polymethacrylate etc. The titrations result in the buildup of negative charge on the polymer and this makes removal of further protons increasingly difficult. In the hydrolysis, the buildup of negative charge on the polymer inhibits attack of hydroxide ions on neighboring carbonyl groups.

In the past the polymer support used in a particular application has, more often than not, been asked to fulfill a number of important functions simultaneously. In this role, the functionalized polymer must possess the correct bulk mechanical properties. Secondly, the support must possess correct porosity and surface area properties. Finally, the support must provide the correct microenvironment to optimize the process being carried out. Increasingly, however, it is being realized that optimum performance can only be achieved by decoupling these functions and, if necessary, optimizing each one independently of the other in terms of micro and macro structure.

1.4.6. Conformation of Polymer Chains

Functional group reactivity can also be affected by the conformation of the polymer chains.[22,23] Whether the polymer chain exists in the form of a helix or coil or whether the coil is a tight or expanded one can influence the accessibility of a functional group to the chemical reagent. These factors are affected by the stereochemistry of the polymer chains as well as the solvents used to carry out these reactions.

1.5. Functional Polymers: The New Generation

Polymer science has a history of symbiosis between academia and industry and is driven by the interplay between fundamental research and practical applications. Polymer scientists are looking for new phenomena, and new combinations of structural elements to meet the increasing demands of material base. The third wave in the functional group and polymer interrelationships in polymer science has been the development of functional polymers. Advances in the capability to correlate chemical structure with polymer properties have placed a great deal of emphasis on molecular design; specific functional groups have been introduced into polymer structures to achieve special combinations of properties.

Functional polymers have as their characteristics the properties of the functional groups and not so much the bulk mechanical properties.[24] These polymers need to be tailor-made with special attention to the type of functional groups and the position of these in the macromolecule which may impart on the polymer chemical, spectral or biological properties characteristic of them. With the current trend to employ polymers as functional materials and not merely as structural materials, the latest trend is to design polymers where the functional group property is the dominant trait. Many of the design considerations originate from "supported chemistry" where the polymer acts as a passive support and not as an active carrier of functional polymer. It would appear that for these functional polymers a science can and should be developed akin to organic functional group chemistry comprising a body of general principles relating on one hand to the molecular structure and properties of the linkage in the material and on the other to its macromolecular properties. Perhaps, the differences between a functional polymer and a functionalized support are not too obvious and need some clarification.

(1) Functional polymers may contain the functional groups in the backbone, side chains or the crosslinks while in all supported chemistry, functional groups on the support can only be as a side chain. This is because the polymer is acting as a carrier for the catalyst or for the reactant.

(2) In the case of functional polymers, the transformation of the functional group is associated with the transformation of the polymer itself. The transformation here may be reversible or irreversible, and may lead to physical changes of the polymer chain. In the case of functionalized supports, seldom do molecular weight changes take place.

(3) The performance of a functional polymer is essentially linked to the chemical property of the functional group rather than the physical characteristics of the polymer.

1.6. Azo Functional Polymers: An Introduction

Functional polymers have become of interest in biology and medicine, in energy production and conservation, in agriculture, in food production and preservation, and in nutrition and health care. Conventional methods and novel approaches are being used to accomplish new designs in macromolecular architecture and to tailor-make multipurpose polymeric structures. Functional groups may impart on the polymer chemical spectral, biological/medicinal properties; they can be photo and thermally or biologically active, or they might have utility as drugs or catalysts.

There has been a proliferation in the variety of uses to which specifically chemically derivatized polymers have been put. Polymer-immobilized drugs,[25] biocides,[26] insect repellents and attractants,[27] fertilizers,[28] food antioxidants and sweeteners,[28] and sunscreens and perfumes[27] have all been investigated. Some of those involve a simple physical encapsulation by a polymer with release by some diffusional mechanism, but many others involve a specific functional group attachment to, and release from, macromolecules. Others involve carefully tailored polymers which themselves degrade within the environment in which they are used.

A number of functional polymers are designed for photochemical applications such as photocrosslinking,[29] solar energy conversion,[30] photodegradation[31] and as photochromic and photoresponsive polymers.[10] Photolabile polymers bearing a variety of functional groups are employed in applications ranging from conventional lithographic techniques to relatively recent stereolithography.[32,33]

Thermally degradable and thermally cross-linkable polymers have been designed by the incorporation of various structural units exhibiting a range of thermal characteristics.[34] Many of these thermolabile polymers have been employed as polyfunctional and sequential initiators and in the synthesis of block and graft copolymers.

Biocompatible surfaces have been designed by the surface modification of polymers involving surface functionalization or grafting of hydrophilic monomers[35,36] by various techniques like radiation grafting and plasma polymerization etc. The area of biodegradable polymers[37,38] has seen phenomenal growth with a number of polymers finding use in drug-delivery systems, surgical sutures, artificial skin, etc.

Another interesting class of polymers which can participate in enzyme-catalyzed side chain reactions has been reported recently.[39,40] If the reaction is found versatile, a variety of optically active polymers bearing side chain asymmetric centers could be prepared by side chain transformations through biocatalysis.

However, very few functional groups in organic chemistry exhibit a range of thermal, chemical, photochemical and biological properties and where each of these properties has been the basis for a distinct class of functional polymers. The azo linkage, the double bonded functionality, is one such group. The present attempt is an outgrowth of our own research interest in the area of biodegradable polymers and photoresponsive polymers based on this functional group. The chemistry of the azo functional group is too massive even for a brief coverage and has been dealt with in detail by a number of authors (Chapter 2). Considerable interest is focussed on the thermal, photochemical and biological properties of these compounds and their utilization as building blocks in the design of functional polymers, but not all the properties have proliferated into macromolecular wonderland. The characteristic features of this group which have become the focus of attention are schematically represented in the diagram below (Figure 1.5).

Although these areas have grown as independent entities, all of them have the same chemical basis in having their origins in the special properties of the azo linkage. There are other properties of this functional group such as transition metal chemistry, photolysis, rearrangements, electrochemical reductions etc., which have not caught much of polymer chemists' imaginations. It is precisely to point out this unified concept of

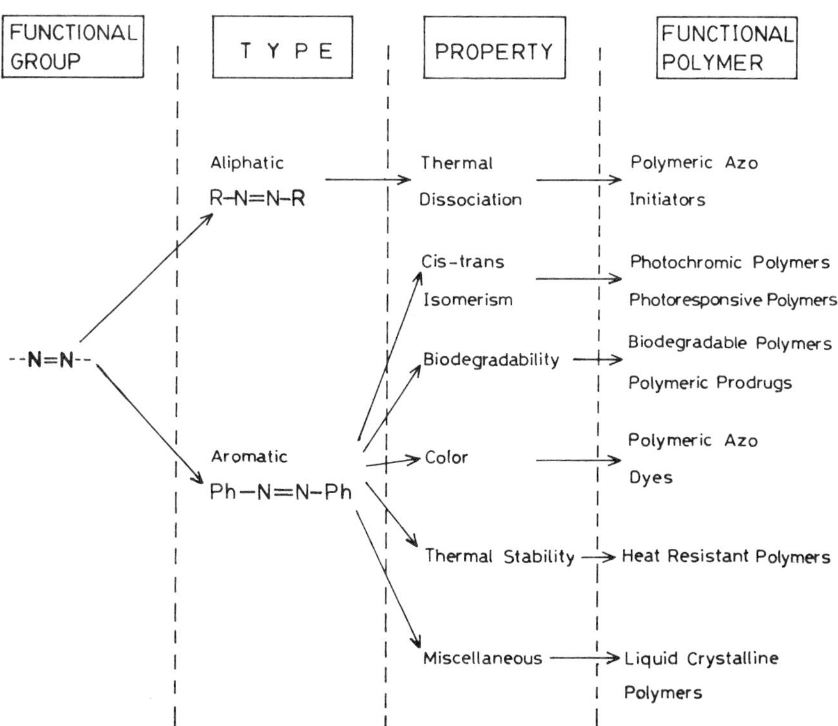

FIGURE 1.5 Functional group approach in azo polymer design.

molecular basis that I have undertaken this monograph. The range of reports is vast and complex. In several instances, the information available is incomplete and uneven. Any classification and summarization must be arbitrary. The general plan of this volume includes the following structure:

- an introductory chapter dealing briefly with the general properties of the azo functional group, especially those properties which have attracted polymer chemists' attention
- a chapter dealing with the strategies in the synthesis of azo functional polymers emphasizing design considerations rather than the detailed methodologies

Subsequent chapters deal with the properties of these azo functional polymers.

- thermal chemistry of azo polymers, both aliphatic and aromatic type leading to thermolabile and thermostable polymers respectively
- photochemistry of azo polymers
- biodegradation of azo polymers and their use as carriers for bioactive agents
- miscellaneous uses as polymeric dyes, liquid crystalline and conducting materials, etc.

This work is not intended to be encyclopedic in nature so I have not mentioned every published azo polymer structure but the treatment is up-to-date. It is a brief critical compilation and owing to the nature of the exercise, it has been necessary to be selective. I have taken the opportunity to discuss "blind alleys" and "dead ends" that exist in the azo functional polymer area.

1.7. References

1. Morawetz, H. 1985. *Polymers: The Origins and Growth of a Science*, New York: John Wiley & Sons.
2. Flory, P. J. 1937. "Kinetics of Condensation Polymerization Reaction of Ethylene Glycol with Succinic Acid," *J. Amer. Chem. Soc.*, 61:1518.
3. Flory, P. J. 1953. *Principles of Polymer Chemistry*, Ithaca: Cornell University Press, p. 102.
4. Merrifield, R. B. 1963. "Solid Phase Peptide Synthesis—I," *J. Amer. Chem. Soc.*, 85:2149.
5. Sherrington, D. C. and P. Hodge, ed. 1988. *Synthesis and Separations Using Functional Polymers*, New York: John Wiley & Sons.
6. Hodge, P. and D. C. Sherrington, ed. 1980. *Polymer-Supported Reactions in Organic Synthesis*, New York: John Wiley & Sons.

7. Mathur, N. K., C. K. Narang and R. E. Williams. 1980. *Polymers as Aids in Organic Chemistry*, New York: Academic Press.
8. Ford, W. T., ed. 1986. *Polymeric Reagents and Catalysts*, ACS Symposium Ser., No. 308. Washington D. C.: ACS.
9. Geckeler, K., V. N. R. Pillai, and M. Mutter. 1981. "Applications of Soluble Polymeric Supports," *Adv. Polym. Sci.* 39:65.
10. Kumar, G. S. and D. C. Neckers. 1989. "Photochemistry of Azobenzene–Containing Polymers," *Chem. Rev.*, 89:1915.
11. Gaylord, N. G. 1968. "Critical Factors Affecting Chemical Reactions on Polymers," *J. Polym. Sci. Polym. Symp.* C(24):1.
12. Overberger, C. G. and B. Sedlacek. 1974. "Transformations of Functional Groups on Polymers.," *J. Polym. Sci. Polym. Symp. Series*, Vol 47:359.
13. Neckers, D. C. 1978. "Solid Phase Synthesis," *Chem. Tech.*, p. 108.
14. Hodge, P. 1988. *Encyclopedia of Polymer Science & Engineering*, New York: John Wiley & Sons. 12:618.
15. Benham, J. L. and J. F. Kinstle. 1988. *Chemical Reactions on Polymers.*, ACS Symposium Ser. 364. Washington D.C.: ACS, p. 1.
16. Parkinson, T. M., J. P. Brown and R. E. Wingard. U.S. Pat. 4,190,716, 1980.
17. Gaonkar, S. R., N. Sapre, S. Bhaduri and G. S. Kumar. 1990. "Polymeric Blocked Isocyanates by the Reductive Carbonylation of Nitropolystyrene," *Macromolecules*, 23:3533.
18. Morawetz, H. 1978. "Comparative Studies of the Reactivity of Polymers and Their Low-Molecular-Weight Analogs," *J. Polym. Sci. Polym. Symp.*, 62:271.
19. Regen, S. L. 1974. "Influence of Solvents on the Mobility of Molecules Covalently Bound to Polystyrene Matrices," *J. Amer. Chem. Soc.*, 96:5275.
20. Ringsdorf, H. 1975. "Structure and Properties of Pharmacologically Active Polymers," *J. Polym. Sci. Polym. Symp.*, 51:135.
21. Vandewijer, P. H. and G. Smets. 1968. "Photochromic Polymers," *J. Polym. Sci.*, C22:251.
22. Kumar, G. S., P. DePra and D. C. Neckers. 1984. "Chelating Polymers Containing Photosensitive Functionalities–I," *Macromolecules*, 17:1912.
23. Kumar, G. S., P. DePra, K. Zhang and D. C. Neckers. 1984. "Chelating Polymers Containing Photosensitive Functionalities–II," *Macromolecules*, 17:2463.
24. Takemoto, K., Y. Inaki and R. M. Ottenbrite. 1988. *Functional Monomers and Polymers*. New York: Marcel Dekker, p. 1.
25. Donaruma, L. G., O. Vogl and R. M. Ottenbrite. 1983. *Polymer Yearbook*. H. G. Elias and R. A. Pethrick, eds. Switzerland: Harwood Church, p. 259.
26. McCormick, C. L., K. W. Anderson and B. H. Hutchinson. 1982/83. "Controlled Activity of Polymers with Pendantly Bound Herbicides," *J. Macromol. Sci. Rev. Macromol. Chem. Phys.*, C(22):57.
27. Campion, D. G., R. Lester and B. F. Nesbitt. 1978. "Controlled Release of Pheromones," *Pestic. Sci.*, 9:434.
28. Akelah, A. 1984. "Biological Applications of Functionalized Polymers," *J. Chem. Tech. Biotech.*, 34(A):263.

29. Guillet, J., ed. 1973. *Polymers and Ecological Problems*, New York: Plenum Press, p. 1.
30. Kaneko, M. and A. Yamada. 1984. "A New Photosensor Composed of Photoresponsive Polymer Membrane," *Adv. Polym. Sci.*, 55:1.
31. Guillet, J. 1985. Chapter 12 in *Polymer Photophysics and Photochemistry*, Cambridge: Cambridge University Press, p. 314.
32. Neckers, D. C. 1990. "An Introduction to Stereolithography," *Chem. Tech.*, p. 615.
33. Kumar, G. S. and D. C. Neckers. 1991. "Laser-Induced Three Dimensional Photopolymerization Using Visible Initiators and UV Crosslinking by Photosensitive Comonomers," *Macromolecules*, 24:4322.
34. Anand, P. S., H. G. Stahl, W. Heitz, G. Weber and L. Battenbach. 1982. "Block Copolymers by Radical Copolymerization," *Makromol. Chem.*, 183:1685.
35. Goldberg, E. P., J. W. Sheets, G. S. Kumar, J. W. Burns and D. C. Osborn. U.S. Pat. 037,153, April 10, 1987.
36. Goldberg, E. P., R. Robinson, A. Ito, J. Stacholy, M. Yalon and G. S. Kumar. 1990. "Surface Graft Polymerization for the Preparation of New Implant and Device Materials," *Polym. Preprin.*, 31:216.
37. Kumar, G. S., V. Kalpagam and U. S. Nandi. 1982/83. "Biodegradable Polymers: Prospects, Problems and Progress," *J. Macromol. Sci. Rev. Macromol. Chem. Phy.*, C22:225.
38. Kumar, G. S. 1986. *Biodegradable Polymers—Prospects and Progress*, New York, NY: Marcel Dekker.
39. Ghogare, A. and G. S. Kumar. 1989. "Oxime Esters as Novel Irreversible Acyl Transfer Agents for Lipase Catalysis in Organic Media," *J. Chem. Soc. Chem. Commun.*, p. 1533.
40. Ghogare, A. and G. S. Kumar. 1990. "Novel Route to Chiral Polymers Involving Biocatalytic Lipase-Catalyzed Transesterification of O-Acryloyl Oximes," *J. Chem. Soc. Chem. Commun.*, p. 135.

2

THE CHEMISTRY OF AZO FUNCTIONAL GROUP

2.1. Introduction

Azo compounds are organo-nitrogen derivatives with the characteristic double-bonded $-N=N-$ functionality and the general formula $R-N=N-R$ where R and R′ can be the same or different groups[1-3] (Figure 2.1). Azo group (**1**) can best be described as a bridging group and the rest of the molecule containing it can be thought of as two moieties, one on either side of the bridge. The group has given its name to a large group of compounds. There are families of organic, inorganic and organometallic azo compounds. Only those structure-property aspects which have had direct bearing in the development of functional polymers are touched here. For example, no mention of the cyclic azo compounds or inorganic azo compounds is made since there is no corresponding polymer literature to justify their inclusion. In this chapter concentration is only on the organic family which has several branches depending on the residues attached to the azo group. The moieties flanking the group are most commonly alkyl or aryl groups. (The historic terminology referring to these compounds as azo compounds has been replaced by the systematic one as derivatives of "diazene." Azobenzene is listed in chemical abstracts as "diazene, diphenyl." The *cis*-isomer is termed as *Z*-isomer and the *trans* as *E*-isomer). However, the terminology in polymer science refers to these compounds as azo derivatives, e.g., as azobenzene and its isomers as *cis* and *trans*. To avoid confusion, I am employing the conventional terminology throughout the book.)

The presence of a double bond generally has profound consequences on the stereochemistry; it gives the molecule more rigidity and restricts the number of possible forms. The main feature is the coplanarity of all atoms adjacent to a double bond as well as of the double bonded nitrogen atoms themselves.[3] The term aliphatic azo compound is taken to cover all those

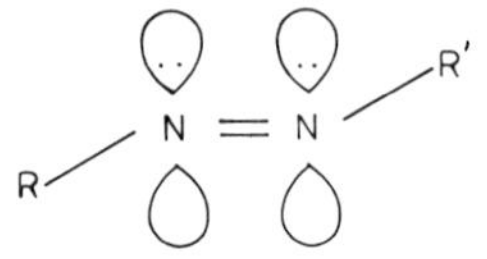

(1)

FIGURE 2.1 The structure of azo functional group. R and R′ = alkyl or aryl groups.

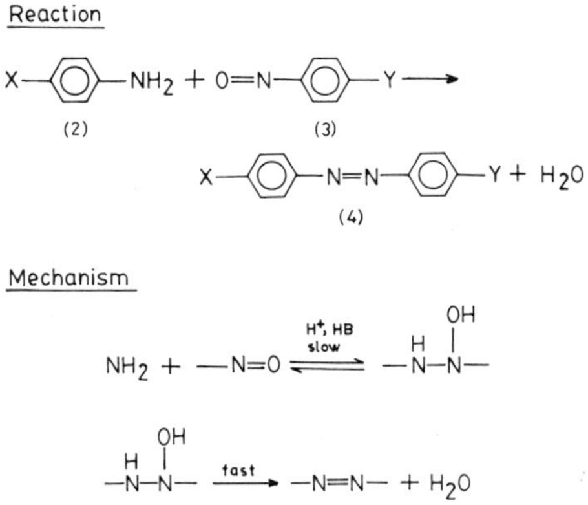

Reaction

(2) (3) (4)

Mechanism

FIGURE 2.2 Condensation reaction of aniline with nitrosobenzene.

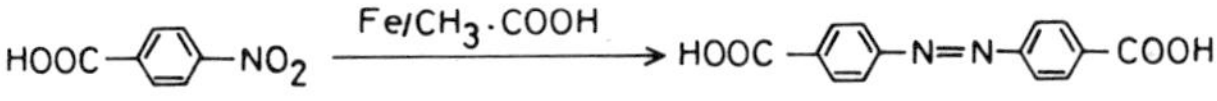

(5) (6)

FIGURE 2.3

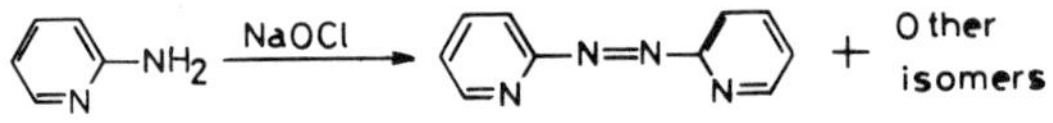

(7) (8)

FIGURE 2.4

substances in which at least one of the two nitrogen atoms is bonded to a sp^3 hybridized carbon atom. Both chemically and technologically, the most important reaction of these aliphatic compounds is their thermolytic fission into nitrogen and free radicals. The bond dissociation reaction can also occur photochemically. The aromatic azo compounds are highly colored explaining the widespread utilization in the dye industry.[4] A brief summary of the properties of the azo group is discussed in this chapter with emphasis only on those properties which have been of particular interest to polymer chemists. The intention is to present a unified discussion of the azo group focussing on molecular properties and lay a foundation for the subsequent chapters which elaborate on how each property has led to a distinct class of functional polymers.

2.2. Methods of Preparation

Azo aromatic linkage is incorporated into the molecules by many different methods, of which the following are the most representative.

2.2.1. Preparation of Aromatic Azo Compounds

A. Condensation Reaction

Ogata and Tagaki[5] following the earlier efforts of others have demonstrated that condensation of anilines(2) with nitrosobenzenes(3) to yield azobenzenes(4) is subject to acid catalysis in acetate-buffered 94% aqueous ethanol solution. Rate constants increased with increasing electron-withdrawing power of polar substituents in the nitrosobenzene moiety and to decrease with increasing electron-withdrawing power of polar substituents in the aniline moiety. These results were interpreted in terms of rate-determining acid-catalyzed addition of amine to the nitroso group followed by rapid dehydration of the resulting intermediate[6] (Figure 2.2).

B. Reductive Coupling

This route involves the reduction of nitrobenzoic acid(5) with iron and acetic acid,[7] with sodium amalgam, alkali sulfides, cellulose, molasses or dextrose in alkaline solution and by catalytic reduction. The reduction with zinc and sodium hydroxide is a modification of Alexejew's method.[7]

C. Oxidative Coupling

Aniline has been oxidized to azobenzene by air and by potassium permanganate.[7] Azopyridines(8) have also been synthesized by the oxidation of the aminopyridines(7) by NaOCl.[8] Similarly phenyl azopyridines are prepared by the oxidation of aminopyridine and aniline.[9,10]

D. Diazotization

The coupling reaction between an aromatic diazo compound(10) and a coupling component(11) is the single most synthetic route to azo dyes.[11-14] Peter Greiss synthesized the first azo dye soon after his discovery of the diazotization reaction in 1858. The two reactions which form the basis for azo dye chemistry are diazotization and coupling as shown in Figure 2.5, where X = Cl, Br, NO_3, HSO_4 etc. All coupling components (11) used to prepare azo dyes possess one common feature i.e., an active hydrogen atom bound to a carbon atom. Compounds of the following type can be used as azo coupling components: compounds that possess phenolic hydroxyl groups such as phenols and naphthols, aromatic amines, compounds that possess enolizable ketone groups of an aliphatic character, and compounds that have an active methylene group. A thorough discussion of diazotization, diazo compounds and azo coupling reactions is found in Saunders,[11] Zollinger,[1] Tedder[13] and Stead.[14]

2.2.2. Preparation of Aliphatic Azo Compounds

Aliphatic compounds containing azo moiety have been prepared by the following two general methods.

A. Oxidation of Hydrazines

Reaction of a ketone(13) with hydrazine to form an azine(14) to which hydrogen cyanide is added to produce an intermediate 1,2 disubstituted hydrazine(15) which is oxidized to the symmetrical azonitrile(16).[15,16]

B. Oxidative Coupling

This approach involves the following sequence of steps: (1) reaction of a ketone with HCN to form a cyanohydrin(17), (2) reaction with ammonia to produce an aminonitrile(18), and (3) oxidative coupling by hypohalite to symmetrical azonitrile(19).[17-24]

Except for simple aminonitriles, the oxidative coupling requires cationic surfactants. Mixtures of symmetrical and unsymmetrical azodinitriles have been prepared by oxidative coupling of a mixture of two different aminonitriles.

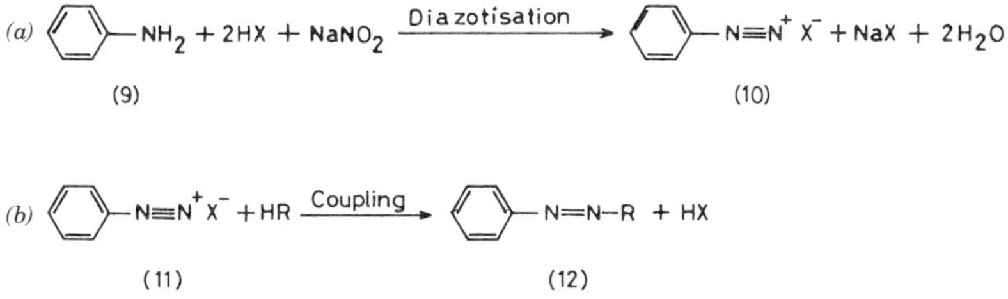

$$(a)\quad C_6H_5{-}NH_2 + 2HX + NaNO_2 \xrightarrow{\text{Diazotisation}} C_6H_5{-}N{\equiv}N^{+}X^{-} + NaX + 2H_2O$$

(9) (10)

$$(b)\quad C_6H_5{-}N{\equiv}N^{+}X^{-} + HR \xrightarrow{\text{Coupling}} C_6H_5{-}N{=}N{-}R + HX$$

(11) (12)

FIGURE 2.5

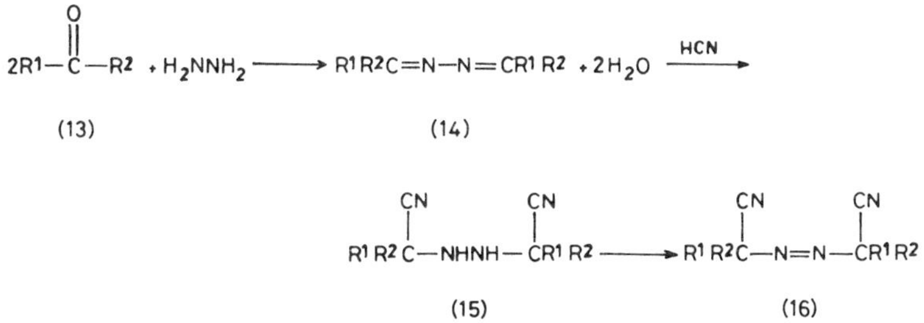

$$2R^1{-}\overset{\overset{O}{\parallel}}{C}{-}R^2 + H_2NNH_2 \longrightarrow R^1R^2C{=}N{-}N{=}CR^1R^2 + 2H_2O \xrightarrow{HCN}$$

(13) (14)

$$R^1R^2\overset{CN}{\overset{|}{C}}{-}NHNH{-}\overset{CN}{\overset{|}{C}}R^1R^2 \longrightarrow R^1R^2\overset{CN}{\overset{|}{C}}{-}N{=}N{-}\overset{CN}{\overset{|}{C}}R^1R^2$$

(15) (16)

FIGURE 2.6

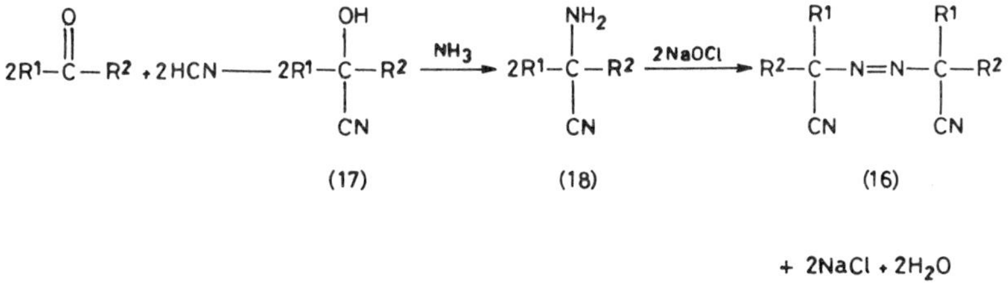

$$2R^1{-}\overset{\overset{O}{\parallel}}{C}{-}R^2 + 2HCN \longrightarrow 2R^1{-}\underset{\underset{CN}{|}}{\overset{\overset{OH}{|}}{C}}{-}R^2 \xrightarrow{NH_3} 2R^1{-}\underset{\underset{CN}{|}}{\overset{\overset{NH_2}{|}}{C}}{-}R^2 \xrightarrow{2NaOCl} R^2{-}\underset{\underset{CN}{|}}{\overset{\overset{R^1}{|}}{C}}{-}N{=}N{-}\underset{\underset{CN}{|}}{\overset{\overset{R^1}{|}}{C}}{-}R^2$$

(17) (18) (16)

$$+ 2NaCl + 2H_2O$$

FIGURE 2.7

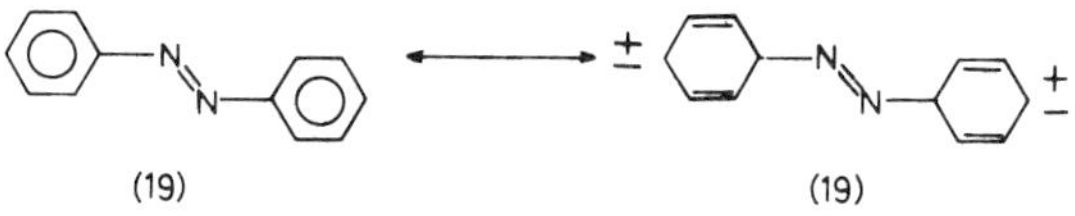

$$C_6H_5{-}N{=}N{-}C_6H_5 \longleftrightarrow \pm\, C_6H_5{-}N{=}N{-}C_6H_5 \,\pm$$

(19) (19)

FIGURE 2.8

2.3. Characterization

Elemental analysis, ultraviolet (UV) spectroscopy, differential scanning calorimetry (DSC) and volumetric analysis are employed to determine the azo content.[25-29] The thermolysis constant k_d is determined by UV, DSC and volumetric methods for aliphatic compounds. The azo group is difficult to analyze by IR spectroscopy.[30] When the azobenzenes are substituted in the para position to the azo group, strong absorption occurs near 1370 cm^{-1} and 1150 cm^{-1}. There are also many bands in compounds of this type in the region between 1615–1000 cm^{-1}.

2.4. Structural Chemistry

Azo compounds are relatively stable compounds and therefore a good number of different structures of azo compounds have been determined.[31,32] Most molecules are planar with a preference for the *trans* form. Bond lengths of 1.22–1.25 Å are found for aliphatic compounds whereas simple aromatic compounds or compounds with conjugated carbonyl groups (with –N=N–) have slightly longer N=N bonds of 1.24–1.26 Å.[32] This indicates some resonance in aromatic compounds. For molecules with several azo groups the N=N bond lengths do increase further but without the corresponding decrease in C–N bond length. The N–N–C bond angles for the aromatic compounds are about 113° and for aliphatic ones about 115°.[33]

Aromatic azo compounds are resonance stabilized. Evidence for this comes from the shortening of the carbon-nitrogen bond in *trans* azobenzene from its usual value of 1.47 Å to 1.41 Å as a result of conjugation. In addition to the *trans* isomer shown, the geometrical *cis* isomer exists which can readily be isolated.[34,35] For azobenzene itself, *trans* form is known to be considerably more stable than the *cis* form, which is directly related to the resonance stabilization in the almost coplanar structure of the former,[36] while in *cis* isomer phenyl rings rotated out of plane containing the nitrogen atoms by about 50°. Although many of the reactions discussed in this chapter are generally valid for all the aromatic azo compounds, the literature disproportionately contains data on azobenzene and its derivatives. Again, the electronic effects i.e., directing and activating effects of phenylazo and arylazo substituents are better known than those of the aliphatic group. The bulk of the quantitative data available on substituent effects of azo groups is in fact derived from the measurements of the effect of meta and para phenyl azo- or arylazo groups on reactions of azobenzene nucleus only.[37] The results of such studies have also been interpreted, where possible in terms of Hammet substituent constants. There are some indications (*vide infra*) that azo containing groups may behave as +R as well as −R groups, depending on the electronic requirement of the reaction. Unfortunately, lack of any information on substituent effects of the alkyl group prevents one from deciding to what extent this flexibility is due to (1) the azo group itself and/or (2) to the interaction between two aromatic systems, with the azo group operating merely as an electron conducting bridge. These molecular characteristics have led to the interest in using azo aromatic compounds in liquid crystalline and conducting polymer applications.

2.5. Properties of Azo Functional Group

2.5.1. Thermal Chemistry

Both chemically and technologically, the most important reaction of aliphatic compounds is their thermolytic fission into molecular nitrogen and free radicals, which subsequently can undergo further processes, such as recombination or transfer reactions (Figure 2.9).

Azo compounds used commercially as free radical initiators contain at least one tertiary carbon atom attached to the azo nitrogen atom. Such azo compounds when appropriately substituted decompose upon heating by homolytic cleavage of two carbon-nitrogen bonds to generate two free radicals and nitrogen. These radicals are reactive intermediates (half lives

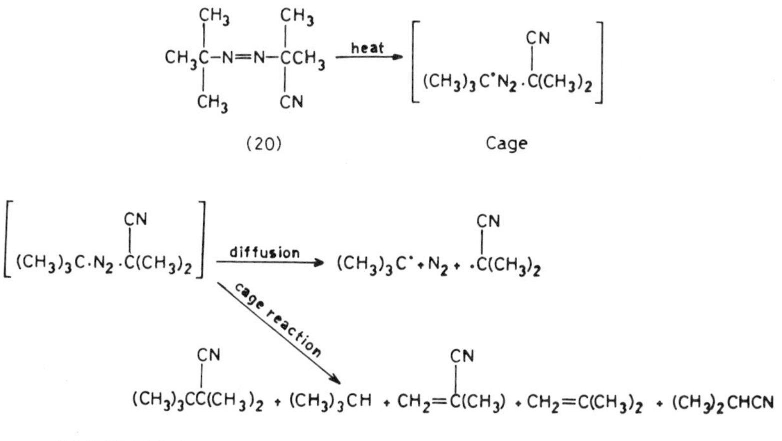

$CH_3C(CH_3)_2{-}N{=}N{-}C(CH_3)_2CN$ (20) $\xrightarrow{\text{heat}}$ $[(CH_3)_3C^{\cdot}N_2{\cdot}C(CN)(CH_3)_2]$ Cage

$[(CH_3)_3C{\cdot}N_2{\cdot}C(CN)(CH_3)_2]$ $\xrightarrow{\text{diffusion}}$ $(CH_3)_3C^{\cdot} + N_2 + {\cdot}C(CN)(CH_3)_2$

cage reaction $\rightarrow$ $(CH_3)_3CC(CN)(CH_3)_2 + (CH_3)_3CH + CH_2{=}C(CN)(CH_3) + CH_2{=}C(CH_3)_2 + (CH_3)_2CHCN$

FIGURE 2.9 Mechanism of decomposition of an azonitrile initiator.

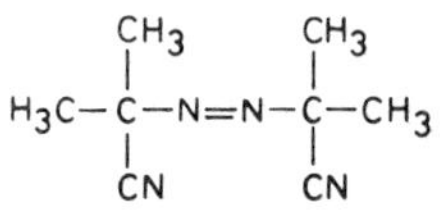

$$H_3C-\underset{CN}{\overset{CH_3}{C}}-N{=}N-\underset{CN}{\overset{CH_3}{C}}-CH_3$$

(21)

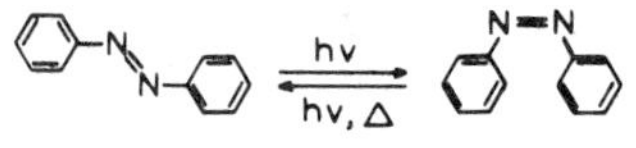

trans (22) $\underset{h\nu,\ \Delta}{\overset{h\nu}{\rightleftharpoons}}$ cis (23)

FIGURE 2.10 Isomerization of azobenzene.

of less than 10^{-3} sec) and have been used to initiate a variety of polymerizations. Although the initiating species is the free radical, the term free radical initiator is used synonymously with the precursor azo compound.

In 1896, Thiele and Heuser, the discoverers of azobisisobutyronitrile(21) (AIBN) now so useful industrially, described the quantitative evolution of nitrogen. The impetus came to study the thermal chemistry of azo compounds in 1949 when it was recognized how suitable AIBN was for the initiation of polymerization. Subsequently many schools conducted research on the decomposition and application of related compounds. Earlier, the peroxides were the most used radical producing agents for polymerization but now azo compounds have achieved about equal importance as the initiators both scientifically and technologically. The experimental investigation of the decomposition of aliphatic azo compounds is relatively simple in comparison with that of other radical reactions. Thermal fission is strictly of the first order; it is affected little by the nature of the solvent and suitable substituents enable the reactivity to be varied within wide limits. A series of investigations have thrown light on the character of the free radicals produced from azo compounds. These compounds exhibit simultaneous homolytic fission of the two links, two radicals being formed as well as a molecule of nitrogen. Whilst in the case of one-bond initiators a high rate of formation of radicals is brought about by the stability of the radicals produced, in the case of azo compounds the rapidity of decomposition is determined by the enormous stability of one of the products—the nitrogen molecule. This permits the formation of relatively unstable radicals, that is, those well suited to initiate chains.

The rate of decomposition of initiators is very important in polymerizations because it exerts considerable influence on the kinetics of the overall process. In the steady state, the rate of polymerization depends according to the equation on the square root of the concentration of the initiator as well as that part of rate of decomposition.

$$\nu_{\text{total}} = -d[M]_{dt} = K_p[fK_i/2K_t]^{1/2}[M][I]^{1/2}$$

where K_i, K_t and K_p are initiation, termination and propagation rate constants; $[M]$ and $[I]$ denote monomer and initiator concentrations respectively.

The formation of radicals from the initiator is followed by addition to a molecule of monomer, but because of the "cage effect," some of the radicals are lost through recombination. Hence not all the molecules of the initiator take part in the initiation and the initiator efficiency (f) is designated as that fraction which successfully reacts with the monomer. The elucidation of the mechanism of chain termination represents an important problem in fundamental technological research in polymerization. Aliphatic azo derivatives are also suitable for photoinitiation of chain reaction since they absorb light in the relatively long wavelength UV region.[3]

2.5.2. Photochemistry

There is a marked difference in the photochemical behavior of aliphatic and aromatic compounds. Aliphatic azo compounds usually undergo photodissociation to give free radicals and nitrogen, where as aromatic azo compounds are stable towards C–N bond scission. Both types may undergo *cis-trans* photoisomerization, although this is not always readily observable. Mixed aliphatic aromatic compounds more closely resemble the pure aliphatic series in their photochemical behavior. The photochemistry of acyclic aliphatic azo compounds has been studied actively for many years, particularly as a clean source of free radicals.[39] Considerable attention is given to aspects related to the mechanism of decomposition, the nature of the excited states involved and the photoisomerization.[40]

The azo group is isosteric with the ethylene group. Two stable configurations can thus be anticipated which have been called *cis* and *trans*. Indeed, in 1937 Hartley[41] isolated *cis* azobenzene and characterized it and in 1964 Hutton and Steel[42] prepared the first aliphatic *cis* azo compound.

A. Electronic Spectra

Azo compounds of the azobenzene type generally are yellow to red. The main feature of their absorption spectra is a relatively weak long

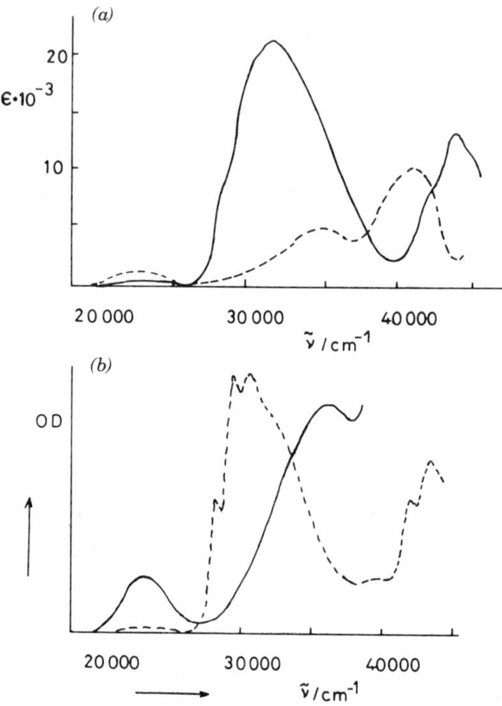

FIGURE 2.11 (a) Absorption spectra of *trans*-azobenzene (—) and *cis*-azobenzene (----) at 298K in EtOH. (b) Absorption spectra of *trans*-azobenzene (—) and *cis*-azobenzenes (----) at 77K in EtOH.

wavelength band well separated from the shorter wavelength band systems.[43-47] Thus the azobenzene type is characterized by a low lying $^1(n-\pi^*)$ state and a large energy gap between this and the next higher $^1(\pi-\pi^*)$ state. Figure 2.11 shows the typical spectra. The low energy $n-\pi^*$ band near 440 nm is without vibrational structure. Indeed, there was never observed any structure on the $n-\pi^*$ band of any non-cyclic azo compound. Compared to the $n-\pi^*$ bands of carbonyl compounds, this azo band is rather intensive. The difference in intensity is due to selection rules, in the planar C_{2h} and C_{2v} symmetries the $n-\pi^*$ is forbidden for the *trans* compound but allowed for the *cis* form. Hence, the intensity of the band in the spectrum of the *trans* form is very high. This has been attributed to the non-planar distortions of the molecule and to vibrational coupling.[47]

The absorption spectra of the shorter wavelength band systems are very similar to those of stilbenes, but the bands are red shifted by about 2000 cm^{-1}. In the spectra of the *trans*-azobenzene these bands are nearly without vibrational structure at room temperature but are clearly structured in a rigid solvent matrix at 77°K. In the spectra of *cis*-azobenzenes the $\pi-\pi^*$ bands generally do not show vibrational features. However, the influence of the molecular structure on the $n-\pi^*$ bands is demonstrated by the spectra of 0–0′ azodiphenylmethane.

The spectra of azoaromatic compounds are rather insensitive to solvent polarity. *Trans*-azobenzene has a $n-\pi^*$ maximum at 442 nm in hydrocarbon and at 437 nm in ethanol solution and p-alkyl, halogen or similar substituents shift both the $n-\pi^*$ and $\pi-\pi^*$ bands.[43,44,48] Substitution of azobenzene by o- or p-amino groups changes the spectra dramatically.[47] The $\pi-\pi^*$ is shifted by several thousand cm^{-1} to longer wavelengths. Thus the compounds of the aminoazobenzene type are characterized by close energy proximity of the $n-\pi^*$ and $\pi-\pi^*$ states. Non-cyclic azo compounds are generally non-emitting and there are very few reports on $n-\pi^*$ emission.

B. Photoisomerization

Azobenzene and many of its derivatives are characterized by reversible transformations from the generally more stable *trans* form to unstable *cis* form upon irradiation with UV or visible light to yield a photostationary composition that is both temperature and wavelength dependent.[48] Azobenzene is thus a textbook representative for demonstrating the rotoresistant property. Many theoretical and experimental studies deal with this subject. The intense absorption at 320 nm due to the absorption of the *trans* isomer decreases upon irradiation while that due to *cis* isomer increases. Thermal isomerism from the photogenerated *cis* form to *trans* form can easily be followed by UV spectroscopy. The thermal isomerism follows first order kinetics and the slope of the plot of log $(A_\infty - A_t)$ against time gives the rate. Quantum yields are generally high for the isomerization of azobenzene and there are no competing reactions of significance. Therefore, the *cis-trans* isomerization represents virtually a model photochemical process in which one isomer is favored photochemically and the other thermally. The conversion of one to the other isomer is quantitative. There is no evidence of known emission from the excited states of azobenzene either in the *trans* or in the *cis* form, so the photochemical process is entirely efficient, at least to the extent that it can be experimentally traced.

Its deaerated hydrocarbon solution can be irradiated with visible or near UV radiation for days without a change of absorption. When oxygen is not excluded a very slow photooxidation to give predominantly azoxybenzene is the only side reaction.[49] But the presence of oxygen does not influence the measured values of isomerization quantum yields. Photoreactions can be analyzed by means of the powerful diagnostics developed by Mauser.[50] Mauser's diagram for the azobenzene isomerization is perfectly linear indicating only one independent reaction in the irradiation system.

For bis azoaromatic compounds, consecutive and simultaneous isomerizations have been reported.[2] Rate data have also been obtained for the *cis-trans* isomerizations of phenylazopyridines and symmetrical azopyridines. Activation energies ranged from 21 to 22.6 k cal/mole^{-1}, with the 2- and 4- substituted compounds showing lower values.[8] Although 2-substituted dyes had low *Ea* terms, the ΔS^* terms were considerably more negative than those of other compounds in the series. The slower isomerization rates for these compounds are attributed to coloumbic interactions between the pyridyl nonbonded electrons and the azo electrons in the activated complex. The 2- and 4- substituted compounds showed a high sensitivity to acid catalysis. Inversions of one of the azo nitrogens has been shown to be overwhelmingly favored over rotation.[8]

One of the interesting issues of azobenzene spectroscopy and photochemistry is the question of mechanism of isomerization. Only a few years after Hartley had found the *trans-cis* isomerization of azobenzene, Magee, Shand and Eyring[51] articulated the idea that there might be an isomerization mechanism that is active which is different from that of stilbene. They proposed a planar transition state with broken double bond in the azo group. The lateral shift mechanism (inversion) concept was introduced by Curtin, Grubbs and McCarthy[52] in 1966 in analogy to the ground state isomerization mechanism of imines. The simple illustration of the two mechanisms is shown in Figure 2.12. Both the mechanisms have their weaknesses; neither of them can rationalize the temperature dependence of the quantum yields of isomerization. There seems to emerge the concept of independent singlet and triplet isomerization pathways. The inversion mechanism in the $^1(n-\pi^*)$ state is gaining support and there is a feeling that the deactivation of the $^1(\pi-\pi^*)$ state at least begins with a rotational motion. Although new experimental materials have been fed into the discussion, the isomerization mechanism is not yet fully understood.

Photoisomerization proceeds with a large structural change as reflected in the dipole moment and change in geometry. The process involves a decrease in the distance between the para carbon atoms in azobenzene from about 9.0 Å in the *trans* form to 5.5 Å in the *cis* form, and the local contraction may even be greater. Likewise *trans* form has no dipole moment while the dipole moment of the non-planar *cis* compound is 3.0 D.[53] Much of the polymer scientists' efforts were directed with the hope that these properties if manifested in the polymers are useful as probes, in

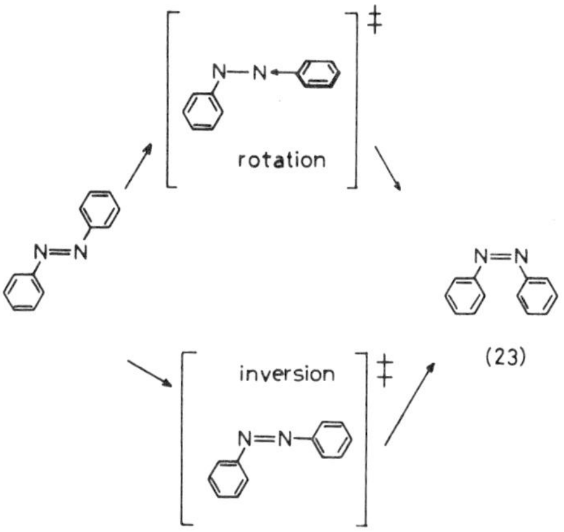

FIGURE 2.12 Rotation and inversion pathways of azobenzene.

conformational dynamics of macromolecules by site-specific labelling, in estimating the free volume of the crosslinked networks, and in designing photoreactive polymers responsive to external stimuli.

The disadvantages of this chromophore as a photosensitive memory system include the following: fast to slow thermal recovery limits the stability of the information deposited and eliminates these compounds as promising candidates for information storage materials, and the fact that not color but just an intensity change constitutes the information.

2.5.3. Color and Constitution

Color, more often than not, is responsible for one's stirrings of interest in chemistry, and it is this visual aspect of the science that makes for much of its appeal to the embryo scientist. The azo dyes are, by far, the most important coloring matter, and in addition to finding widespread use in the coloration of all types of fibers, they provide many useful pigments, histological stains, and analytical colorimetric reagents.[54] The versatility of this class stems largely from the ease with which azo compounds can be made, and almost any diazotized aromatic amine can be coupled with any stable nucleophilic unsaturated system to give a colored azo product (Section 2.2.1.). If the resultant compound contains a primary amino group (24), this may in turn be diazotized and coupled, giving a system of greater conjugation (25) as shown in Figure 2.13.

By extending the conjugation this way, or by incorporating larger ring systems, or different electron donor groups, a full spectral range of colors

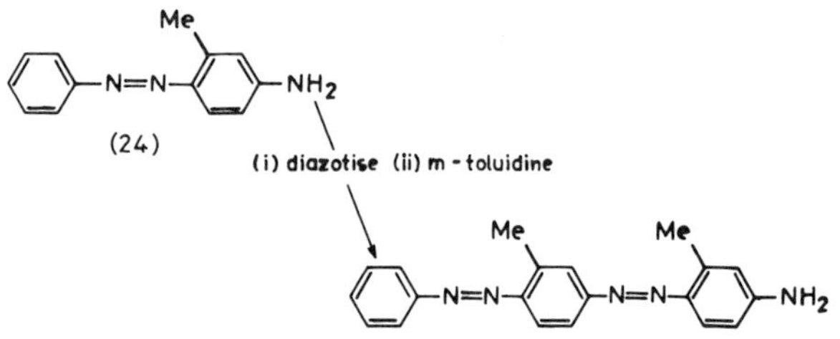

(24)

(i) diazotise (ii) m - toluidine

(25)

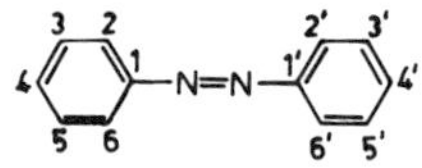

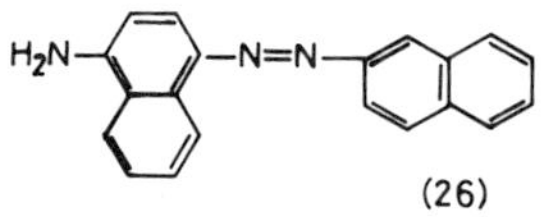

(26)

λ_{max}^{EtOH} 465nm

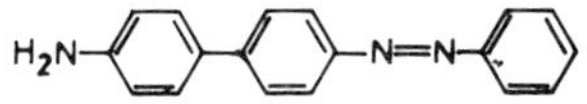

(27)

λ_{max}^{EtOH} 450nm

FIGURE 2.13

can be obtained, with almost any desired chemical or physical properties. In the absence of electron donor groups, azo compounds are only weakly colored, and the visible absorption band corresponds to the low intensity n–π^* transition of the azo group. If an electron donating group is introduced into phenyl rings, then a new intense absorption band arises, usually in the visible region, which is associated with the transfer of electron density from the donor group into the rest of the chromogen. It is useful to designate intensely colored compounds of this type as azo dyes, in order to distinguish them from the weakly colored azo compounds, the latter being devoid of electron donating groups and possessing visible n–π^* bands only. The polarity and thus the absorption wavelength can be increased further by attaching electron-withdrawing groups to the second phenyl ring. Examination of a large number of commercial dyes of this type shows that one unit is invariably a carbocyclic or a heterocyclic unsaturated system, and it may possess from one to three electron releasing groups. The second phenyl ring systems can be similar, with or without electron withdrawing substituents. In Table 2.1, the visible absorption maxima of several donor-substituted azobenzenes are given,[55,56] and it is clear that for a monosubstituted dye, the bathochromic shift depends on the electron donating strength of the substituent. Since the electronic transition in these compounds involves a general migration of electron density from the donor group towards the azo group, it is not surprising that electron-withdrawing groups in the second benzene ring exert a bathochromic shift. The largest shift is observed for the p-NO_2 group, and the smallest shift for halogen atoms. It is found that the bathochromic shift in a given series of dyes is related roughly to the Hammett sigma constant for the electron with the drawing group, but the correlations are not perfect. Increasing the conjugation of an azo dye has the usual bathochromic effect on the visible absorption band, and this may be brought about by increased lateral conjugation (26), usually by including a naphthalene(27) rather than a benzene ring or increased longitudinal conjugation, usually by increasing the number of azo groups (26). (Figure 2.14). It was found that longitudinal extension of the chromogen as in Figure 2.14 is the least effective, and the bathochromic effect accompanying multiplication of the number of azo groups converges rapidly (25,28).

Table 2.1 Visible Absorption Spectra of Some Donor-Acceptor Substituted Azobenzenes.

Substituents	$\lambda_{max}^{(nm)}$ (log ϵ)	Substituents	$\lambda_{max}^{(nm)}$ (log ϵ)
None	318 (4.33)	4-NO_2	332 (4.38)
2-NH_2	417 (3.8)	4-OH-4′-NO_2	386 (4.47)
3-NH_2	417 (3.1)	4-NMe_2-4′-NO_2	478 (4.52)
4-NH_2	385 (4.39)	4-NEt_2-4′-NO_2	490 (4.56)
4-OH	349 (4.42)	4-NMe_2-2′-NO_2	440 (4.43)
4-SMe	362 (4.38)	4-NMe_2-3′-NO_2	431 (4.46)
4-NHAc	347 (4.37)	4-NMe_2-4′-Ac	447 (4.50)
4-NHMe	402 (4.41)	4-NEt_2-4′-Ac	462 (4.45)
4-NMe_2	408 (4.44)	4-NEt_2-4′-CN	466 (4.51)
4-NEt_2	415 (4.47)	4-NEt_2-3′-CN	446 (4.45)
2,4-diNH_{2g}v	411 (4.32)	4-NEt_2-2′-CN	462 (4.48)
2,4-di-O^-	473 (—)	4-NEt_2-2′,4′-diCN	515 (4.60)
3,4-di-O^-	501 (—)	4-NEt_2-2′,6′-diCN	503 (4.52)
2,5-di-O^-	572 (—)	4-NEt_2-3′,5′-diCN	478 (4.53)
2,4-di-O^--4′-NO_2	574 (—)	4-NEt_2-2′,5′-diCN	495 (4.56)
3,4-di-O^--4′-NO_2	613 (—)	4-NEt_2-3′,4′-diCN	500 (4.59)
2,5-di-O^--4′NO_2	655 (—)	4-NEt_2-2′,4′,6′-diCN	562 (4.67)

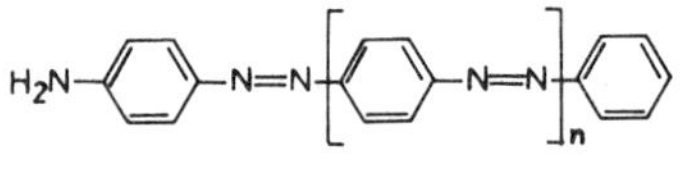

$H_2N-C_6H_4-N{=}N-[C_6H_4-N{=}N]_n-C_6H_5$

(28)

$n=1, \lambda_{max}^{C_6H_6}$ 416nm

$n=2, \lambda_{max}^{C_6H_6}$ 428nm

FIGURE 2.14

Recently, a novel bichromorphoric dye system involving azo benzene units has been reported by Kumar and Neckers.[57] The approach involves coupling two xanthene dye (Rosebengal) moieties through the para position of the azobenzene nucleus as shown in Figure 2.15. The azo-bridged rosebengal dimer(29) exhibits interesting spectral and bleaching characteristics. Upon irradiation at 325 nm, the absorption maxima of the azobenzene moiety, the azo-RB did not undergo photochemical isomerization but instead underwent visible-light induced bleaching of the xanthene dye unlike the analogous rosebengal benzyl ester indicating perhaps some type of energy transfer from the azobenzene nucleus. The extrapolation of this work to polymeric networks might result in some interesting polymeric dyes.

2.5.4. Oxidation

The oxidation reaction of aliphatic and aromatic azo compounds has attracted considerable interest apparently because of their utility for the synthesis azoxy derivatives, which are useful starting materials for a number of reactions.[58-60] Sandler and Karo[61] have critically reviewed the available methods of synthesis including oxidative routes, for both aliphatic and aromatic compounds. The oxidation reactions of azo dyes have also been reviewed elsewhere.[62] Both *trans* and *cis* aliphatic azo

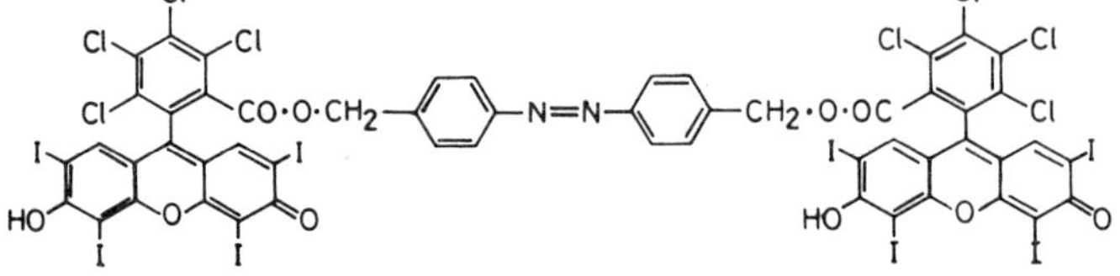

Cl Cl Cl Cl CO·O·CH_2 N=N CH_2·O·OC Cl Cl Cl Cl I I HO O O I I I I HO O O I I

(29)

FIGURE 2.15 Structure of azobenzene-bridged rosebengal dimer.

$$-N{=}N- \xrightarrow{\text{Oxidation}} -N{=}\overset{\overset{O}{\uparrow}}{N}-$$

FIGURE 2.16

compounds may be converted to the corresponding azoxy compounds by the application of appropriate oxidants but since they are prone to acid catalyzed isomerization to hydrazones and are sensitive to decomposition,[63] the reaction conditions have to be controlled to achieve successful conversion. The preferred reagents for the transformations are relatively weak peracids, such as perbenzoic and peracetic acids, since their use in conjunction with suitable solvent diminishes the possibility of isomerization.

Azobenzenes are generally much more stable than aliphatic azo derivatives and this factor has permitted and indeed in many cases has necessitated the use of strong oxidants and drastic conditions to bring about desired conversions to azoxybenzenes. It has been shown that the oxidation of substituted azobenzenes with perbenzoic acid exhibits a second order kinetics and that substituents have a considerable effect on the rate of reaction and the energy of activation.[64] Electron-releasing groups increase the rate while electron attracting substituents decrease the rate of reaction. Badger et al.[65] have determined the rate of reaction of azobenzene with perbenzoic acid in chloroform, and brought to light some interesting differences in the reactivity of *trans* and *cis* isomers. The oxidation of azobenzene at room temperature and in the dark gave *cis* azoxy isomer and that of a *trans* form yielded *trans* azoxy isomer when carried out in the dark. The kinetic study revealed that *cis* azobenzene reacted much more rapidly than that of *trans* isomer, thus indicating that the electron density or electron availability at the azo linkage in the former is considerably larger than in the latter. The report on the oxidation of unsymmetrically substituted aliphatic and aromatic compounds have brought to light some of the factors that determine which of these nitrogens of the azo groups will be oxidized. Oxidation reaction has also been utilized for the oxidation of polymeric azo compounds to azoxy-containing polymers which have received extensive attention as liquid crystalline materials.

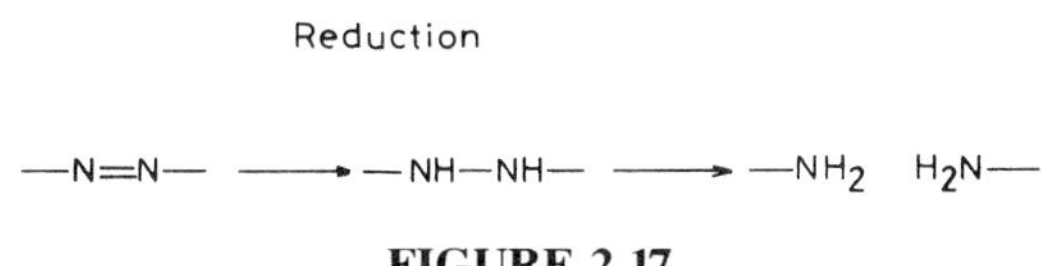

FIGURE 2.17

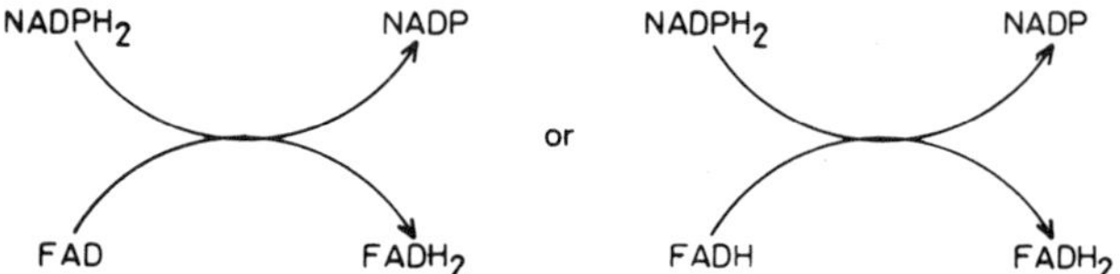

FIGURE 2.18 System postulated for non-specific azo reduction by $FADH_2$.

2.5.5. *Reduction*

The majority of recently reported investigations on the reduction of hydrazo, azo and azoxy compounds have been concerned with azobenzenes and of these by far, more studies have been carried out on symmetrically substituted compounds than on unsymmetrical ones.[66] There has been little interest in the reduction of other azoarenes. Also, little attention has been paid to the reduction of aliphatic azo compounds to the corresponding hydrazo compounds, for which carefully controlled reaction conditions and mild reagents are usually essential.

Many agents have been employed for reductions of azo and related compounds, but the preferred means seem to be lithium aluminum hydride, zinc dust in alkaline or acidic medium, and hydrogenation – the growth in the use of latter has generally been remarkable. The effect of substituents on the course of the reaction has also been touched on in some investigations, particularly with regard to azobenzenes, however more systematic studies are required to fully understand the factors involved.

2.5.6. *Biodegradation*

The *in vivo* azo reduction with cleavage of aromatic azo linkage is one of the earliest explored biological reactions of the functional groups.[67] The two-stage azo reduction proceeding by way of hydrazo intermediates can be regarded as the reverse reaction of non-enzymic and enzymic azo formation from hydrazo and aromatic amines. Azo compounds are mainly reduced in the liver and large intestine.[68]

A. Hepatic Azo Reducing System

The existence of a mammalian enzyme system capable of reducing azo groups was demonstrated by Mueller and Miller.[69,70] Working with butter yellow, these investigators showed that reductive fission of the azo group occurred when the dye was incubated with rat-liver homogenates under anaerobic conditions, that the enzyme responsible was located mainly in the microsomal fraction of the liver and NADPH was the required electron donor. Their work also indicated that the enzyme was probably a flavoprotein with FAD as the prosthetic group and this led to the suggestion that it might be a $NADPH_2$-cytochrome C reductase. It is now generally believed that there exists in the liver an indirect reduction system in which the coenzyme FAD is not a tightly bound prosthetic group but, in its fully reduced form, reduces various substrates non-specifically. The system postulated is similar to that suggested for the reduction of nitro compounds and is shown in Figure 2.18. The studies on hepatic azo reductase have indicated that this enzyme possesses a low substrate specificity since a variety of azo compounds have been used as substrates.

B. Intestinal Azo Reducing Systems

The existence of a hepatic reducing system is not the only site at which considerable azo-reducing activity has been observed *in vitro*.[68] A major advance in the understanding of the metabolism of water-soluble dyes resulted from the work of Radomski and Mellinger[71] who demonstrated that the observed reductive fission of the azo group in the gut was probably due to the action of the gut microflora. A more comprehensive survey of the reduction of azo compounds by microorganisms was carried out by Dickhues (1960) who screened 21 species, including some normal components of the mammalian gut population, against 14 azo compounds and showed that azo reducing activity was fairly general and relatively non-specific.[72] Reduction is without doubt the most important modification that azo compounds undergo in the gut, but other significant metabolic changes do occur. However, the other modifications that can occur in the gut have not been studied extensively. No reports of enzymatic degradation of aliphatic azo compounds are available so far.

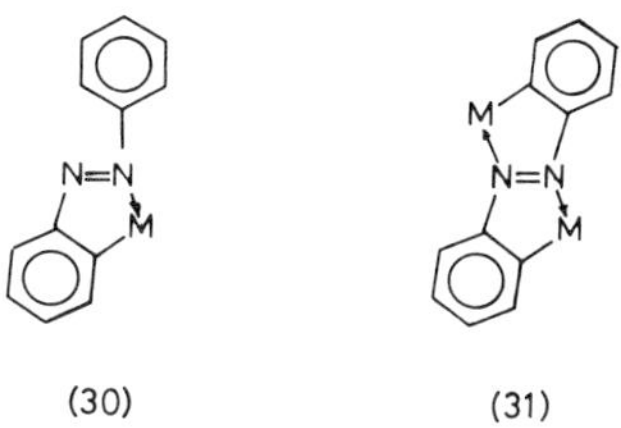

FIGURE 2.19

2.5.7. *Transition Metal Chemistry*

The interaction of azo compounds with transition metal derivatives has produced complexes of at least four different types.[73]

(1) Containing the azo compound attached via nitrogen sigma donor bonds
(2) Containing the azo compound bonded via π-bonds involving electrons of the N=N system
(3) In the case of aromatic derivatives, containing a metallated ligand attached via one nitrogen atom and a metal carbon bond to the ortho carbon atom of the ring (30,31)
(4) Containing rearranged nitrogen-donor ligands, e.g., o-semidines

In general, an important reaction in the chemistry of ligands containing aromatic azo groups is *ortho* metallation, whereby organometallic compounds containing arylcarbon metal sigma bonds are formed, intramolecularly, via displacement of an ortho hydrogen of the aromatic ring by a transition metal atom in the complex. The o-metallation reaction involving azobenzene is one of the most extensively studied. In the resulting 2-(phenylazo)phenyl metal derivatives the ligand is chelated to the metal via a nitrogen donor bond and a metal carbon sigma bond forming a five-membered ring. It is evident that the transition metal chemistry of azobenzene is both extensive and varied, and it is not possible to do more than indicate its ramifications. Future developments in the organometallic field will no doubt result in the elucidation of more precise mechanisms of the o-metallation and the rearrangement reactions where as the organic chemists are likely to find reactions of the resulting complexes to be more exciting. There are no published reports on the o-metallation reactions of polymeric azo compounds.

2.5.8. *Miscellaneous Properties*

A. Non-Linear Optics

When laser radiation passes through a dielectric medium, the intense electromagnetic fields interact with the materials to give rise to new fields altered in frequency, phase or amplitude relative to the incident radiation. This is the basic science of non-linear optics. Current efforts in this area are aimed at materials exhibiting second order non-linear optical phenomena, such as second harmonic generation (SHG) where light of

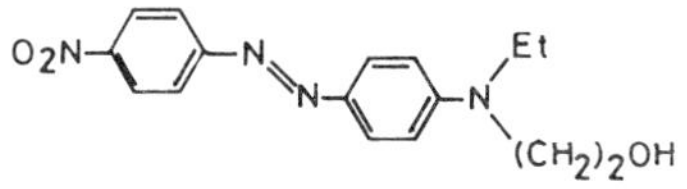

(32)

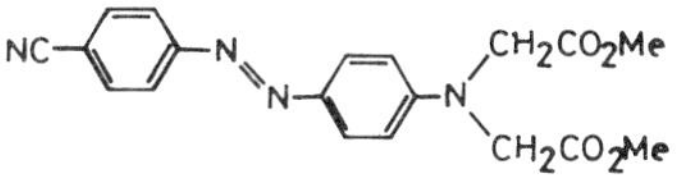

(33)

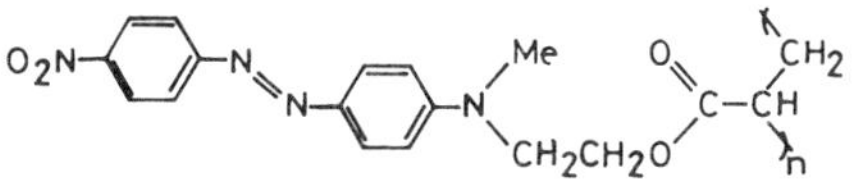

(34)

FIGURE 2.20

frequency *w* is converted to light of frequency 2*w*, and the linear electro-optic effect, whereby the refractive index varies linearly with an applied electric field. For a material to exhibit second order effects, its constituent molecules must possess a finite molecular non-linearity, and must be arranged non-centrosymmetrically in the bulk material, so that their individual non-linearities do not cancel. Early work on electro-optical polymers used active monomeric guests in a polymeric host. One of the first reports of such a system is of the azo dye,[31,32] disperse red 1 in a matrix of PMMA in which second harmonic generation was observed.[74,75] Currently, efforts are underway to design polymeric structures which exhibit a non-linear effect themselves.[34]

B. Chemical Data Storage

Recently, a novel, reversible chemical method involving azobenzene for information storage with ultra high density and nondestructive information read out has been demonstrated by a trio of researchers at the University of Tokyo.[76] It has the potential to be used in optical memory devices. The method makes use of the properties of a film of an azobenzene derivative that can be switched between three chemical states photochemically or electrochemically. The azo compound can be switched from *trans* (35) to *cis* (36) by ultraviolet radiation, with the reaction occurring either ther-

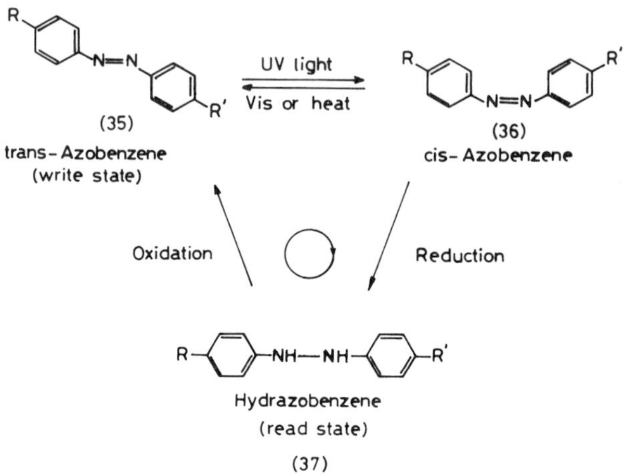

FIGURE 2.21 Azobenzene derivative cycles between three states.

mally or at visible wavelengths. The less stable *cis* isomer may be electrochemically reduced to hydrazobenzene(37) (the read state). To complete the cycle, the hydrazobenzene can be electrochemically oxidized to the *trans* azobenzene. A device based on their chemical system, Tokyo researchers say, would have a storage capacity of 100 million bits per sq cm. This could be increased, they note, if several azobenzene species with different spectral responses were located in the storage film and a turntable laser was used. Fujishima's results are reported for 4-octyl-4′-(5-carboxy-pentamethylene-oxy)-azobenzene, although Fujishima's group has also studied the photoelectrochemical behavior of other long chain fatty acid derivatives.

C. Electrophilic Substitution Reactions

The most commonly encountered azo compounds are those in which both the groups are aryl and as far as electrophilic substitution is concerned, the available information is restricted to scattered data on products isolated as a result of substitution reactions carried out on azobenzene or on simple derivatives thereof. Although nitration, halogenation and sulphonation of azobenzene can be successful, it does appear that azobenzene is resistant to electrophilic attack on the aromatic ring. Most of the other electrophiles are weak and the lack of reactivity to Friedal-Crafts reagents lends support to the assumption that azobenzene behaves as a deactivated substrate towards electrophiles. It is suggested that these reagents might bind to the azo group, by far the most electronegative part of the molecule.[70]

2.6. Conclusion

I have made an attempt to summarize some of the important characteristics of azo functionality with a special emphasis on those which have branched into the domain of polymer science. Thermal, photochemical and biological properties have attracted polymer chemists' attention only in recent years, while transition metal chemistry and rearrangements involving azo moieties have not been investigated on macromolecular substrates. Some of the studies involving chemical data storage and azobenzene bridged dye systems are sure to find applications in the functional polymer area.

2.7. References

1. Zollinger, H. 1961. *Azo and Diazo Chemistry*, New York, NY: Wiley, p. 1
2. Engel, P. S. 1980. "Mechanism of the Thermal and Photochemical Decomposition of Azoalkanes," *Chem. Rev.*, 80:99.
3. Patai, S., ed. 1975. *The Chemistry of the Hydrazo, Azo and Azoxy Groups*, Part I & II, New York, NY: Wiley.
4. Zollinger, H. 1987. *Color Chemistry: Synthesis, Properties and Applications of Organic Dyes*, Weinheim: VCH, p. 1
5. Ogata, Y. and Y. Takagi. 1958. "Kinetics of the Condensation of Anilines with Nitrosobenzenes to Form Azobenzenes," *J. Amer. Chem. Soc.*, 80:3591.
6. Runes, R. A., A. J. Terenzani and L. Amaral. 1975. "Kinetics and Mechanism of Azobenzene Formation," *J. Amer. Chem. Soc.*, 97(2):368.
7. Horning, E. C., ed. 1955. *Organic Synthesis*, New York, NY: Wiley, 3:103.
8. Brown, E. V. and G. R. Granneman. 1975. "Cis-Trans Isomerism in the Analogs of Azobenzene," *J. Amer. Chem. Soc.*, 97(3):621.
9. Campbell, N., A. W. Henderson and D. Taylor. 1953. "Geometrical Isomerism of Azo Compounds," *J. Chem. Soc.*, p. 1281.
10. Brown, E. V. 1969. "Mass Spectra of Some Phenyl Azopyridines and Quinolines," *J. Heterocyclic. Chem.*, 6:571.
11. Saunders, K. H. 1949. *The Aromatic Diazo Compounds, 2nd Ed.*, London: Edward Arnold Co., p. 1.
12. Kirk-Othmer. 1975. *Encyclopedia of Chemical Technology 3rd Ed.*, New York, NY: John Wiley & Sons, 3:387.
13. Tedder, J. M. and K. Venkatraman, eds. 1970. *The Chemistry of Synthetic Dyes*, New York, NY: Academic Press, 3:223–248.

14. Stead, C. V. 1970. In *The Chemistry of Synthetic Dyes*, K. Venkatraman, ed., New York, NY: Academic Press, 3:249–301.
15. Koyanagi, S. et al. Ger. Offen, 1,954,084, 1971.
16. Robertson, A. U.S. Pat. 2,520,338 (to E. I. du Pont de Nemours & Co. Inc.), 1950.
17. Nield, C. H., J. Clarke and C. G. Screttas. U.S. Pat. 3,390,146, 1968.
18. Fuchs, J. U.S. Pat. 3,783,148, 1974.
19. Anderson, A. W. U.S. Pat. 2,711,405, 1955.
20. DeBenneville, P. L. U.S. Pat. 2,713,576, 1955.
21. Moore, E. P. U.S. Pat. 4,028,334, 1977.
22. Moore, E. P. U.S. Pat. 4,051,124, 1977.
23. Moore, E. P. U.K. Pat. 2,099,426, 1982.
24. Walz, R. and W. Heitz. 1978. "Preparation of $(AB)_n$ Type Block Copolymers by Use of Polyazoesters," *J. Polym. Sci. Polym. Chem. Ed.*, 16:1807.
25. Walz, R., B. Bomer and W. Heitz. 1977. "Monomeric and Polymeric Initiators," *Makromol. Chem.*, 178:2527.
26. Oppenheimer, C. and W. Heitz. 1982. "Synthesis and Characterization of Surface-active Azo-group Containing Polyacrylamides," *Angew. Makromol. Chem.*, 98:167.
27. Dicke, H. R. and W. Heitz. 1981. "Synthesis and Characterization of Azo-Group Containing Polyacrylamides—II" *Makromol. Chem. Rapid. Commun.*, 2:83.
28. Dicke, H. R. and W. Heitz. 1982. "Surface-active Azo Group-Containing Polymers in Emulsion Polymerization," *Colloid. Polym. Sci.*, 260:3.
29. Anand, P. S., H. G. Stahl, W. Heitz, G. Weber and L. Bottenbruch. 1982. "Block Copolymers by Radical Copolymerization," *Makromol. Chemie*, 183:1685.
30. Puchert, C. J., ed. 1975. *The Aldrich Library of Infrared Spectra, 2nd Ed.*, Milwaukee: Aldrich, p. 1261.
31. Allmann, R. and S. Patai, eds. 1975. *The Chemistry of Hydrazo, Azo and Azoxy Groups*, Part 1, New York, NY: John Wiley & Sons, p. 23.
32. Mostad, A. and C. Romming. 1971. "The Crystal Structure of p-benzene bis Diazonium Tetrachloro Zincate," *Acta Chem. Scandina.*, 25:3561.
33. Hope, H. and D. Victor. 1969. "Structure of Trans-p, p′Dichloroazobenzene," *Acta Cryst.*, B25:1849.
34. Lange, J. J., J. M. Robertson and I. Woodward. 1939. "X-Ray Crystal Analysis of Trans-Azobenzene," *Proc. Roy. Soc.*, A171:398.
35. Hampson, C. C., J. M. Robertson. 1941. "Bond Length and Resonance in the Cis-Azobenzene Molecule," *J. Chem. Soc.*, p. 409.
36. Happer, D. A. R. and J. Vaughan. 1975. *The Chemistry of Hydrazo, Azo and Azoxy Compounds*, Part 1, S. Patai, ed., New York, NY: John Wiley & Sons, p. 225.
37. Van Meter, J. P. 1977. *The Chemistry of Double-Bonded Functional Groups*, Part 1, S. Patai, ed., New York, NY: John Wiley & Sons, p. 93.
38. Sheppard, C. S. 1985. In *Encyclopedia of Polymer Science & Engineering*, Vol. 2, New York, NY: Wiley & Sons, p. 143.
39. Overberger, C. G., J. P. Anselme and J. G. Lombardino. 1966. Chapter 4 in *Or-*

ganic Compounds with Nitrogen-Nitrogen Bonds, New York, NY: The Ronald Press Co.

40. Smith, P. A. S. 1966. Chapter 2 in *The Chemistry of Open-Chain Organic Nitrogen Compounds*, Vol 2, New York, NY: W. A. Benjamin, Inc.
41. Hartley, G. S. 1937. "Cis-form of Azobenzene," *Nature*, 140:281.
42. Hutton, R. F. and C. Steel. 1964. "Photoisomerization of Azomethane," *J. Amer. Chem. Soc.*, 86:745.
43. Suzuki, H. 1967. *Electronic Absorption Spectra and Geometry of Organic Molecules*, New York, NY: Academic Press.
44. Jaffe, H. H. and M. Orchin. 1962. *Theory and Application of Ultraviolet Spectroscopy*, New York, NY: Wiley.
45. Birnbaum, P. P., J. H. Linford and D. W. G. Style. 1953. "Absorption Spectra of Azobenzene Derivatives," *Trans. Far. Soc.*, 49:735.
46. Gore, P. H. and O. H. Wheeler. 1961. "Absorption Spectra of Aromatic Azo and Related Compounds, Substituted Azobenzenes and Benzanils," *J. Org. Chem.*, 26:3295.
47. Rau, H. 1990. In *Photochromism*, Amsterdam: Elsevier, p. 165.
48. Griffiths, J. 1972. "Photochemistry of Azobenzene and Its Derivatives," *Chem. Soc. Rev.*, 1:489.
49. Ross, D. L. and J. Blanc. 1971. In *Photochromism*, G. H. Brown, ed., New York, NY: Wiley.
50. Mauser, H. Z. 1968. "Spectroscopic Study of Chemical Reaction Kinetics II. Extinction Difference Diagrams," *Naturforsch.*, 23B:1025.
51. Magee, J. L., W. Shand and H. Eyring. 1941. "Non-adiabatic Reactions-Rotations around a Double Bond," *J. Amer. Chem. Soc.*, 63:677.
52. Curtin, D. Y., E. J. Grubbs and C. G. McCarthy. 1966. "Uncatalyzed Syn-Anti Isomerization of Imines, Oxime Ethers and Haloimines," *J. Amer. Chem. Soc.*, 88:2775.
53. Kumar, G. S. and D. C. Neckers. 1989. "Photochemistry of Azobenzene-Containing Polymers," *Chem. Rev.*, 89:1915.
54. 1976. *Color and Constitution of Organic Molecules*, New York, NY: Academic Press, p. 180.
55. Morris, R. J., F. R. Jensen and T. R. Luewebrink. 1954. "Relation between the Absorption Spectra and the Chemical Constitution of Dyes," *J. Org. Chem.*, 19:1306.
56. Sawicki, E. 1957. "Physical Properties of Aminoazobenzene Dyes," *J. Org. Chem.*, 22:915.
57. Kumar, G. S. and D. C. Neckers. In press. "Photochemistry of Azobenzene-Bridged Rosebengal," *J. Photochem. Photobiol.*
58. Stewart, R. 1964. *Oxidation Mechanisms–Applications to Organic Chemistry*, New York, NY: W. J. Benjamin.
59. Wiberg, K. B., ed. 1965. *Oxidation in Organic Chemistry*, New York, NY: Academic Press Inc.
60. Rinehart, K. L. 1973. *Oxidation and Reduction of Organic Compounds*, Englewood Cliffs: Prentice-Hall.

61. Sandler, S. R. and W. Karo. 1971. In *Organic Functional Group Preparations*, Vol 2, A. T. Blomquist, ed., New York, NY: Academic Press.
62. Rys, P. and H. Zollinger. 1972. *Fundamentals of the Chemistry and Applications of Dyes*, London: Wiley, p. 15.
63. Newbold, B. T. 1965. "Preparation and Reactions of 2,2′ Diiodo azobenzene," *J. Chem. Soc.*, p. 6972.
64. Pentimalli, L. 1959. "Aromatic Azo Compounds—Oxidation of 4-dimethyl Aminoazobenzene," *Tetrahedron*, 5:27.
65. Badger, G. M. and G. E. Lewis. 1953. "Aromatic Azo Compounds—Oxidation of Cis and Trans Azobenzenes," *J. Chem. Soc.*, p. 2151.
66. Newbold, P. T. 1975. *The Chemistry of Hydrazo, Azo and Azoxy Compounds*, Part 1, S. Patai, ed., New York, NY: Wiley, p. 541.
67. Miyadera, T. 1975. *The Chemistry of Hydrazo, Azo and Azoxy Groups*, Part 1, S. Patai, ed., New York, NY: Wiley, p. 495
68. Walker, R. 1970. "Metabolism of Azo Compounds: A Review of Literature," *Food Cosmet. Toxicol.* 8:659–676.
69. Mueller, G. C. and J. A. Miller. 1948. "Metabolism of 4-dimethyl Azobenzene by Liver Homogenates," *J. Biol. Chem.*, 176:535.
70. Mueller, G. C. and J.A. Miller. 1949. "Reductive Cleavage of 4-dimethyl Aminoazobenzene by Rat Liver," *J. Biol. Chem.*, 180:1125.
71. Radomski, J. L. and T. J. Mellinger. 1962. "Absorption, Fate, and Excretion in Rats of Soluble Azo Dyes," *J. Pharm. Exp. Ther.*, 136:259.
72. Dieckhues, B. 1960. "Reductive Splitting of Azo Dyes by Bacteria," *Zentl. Bakt. Parasitkde.*, 180:244.
73. Bruce, M.I. and B. L. Goodall. 1975. In *The Chemistry of Hydrazo, Azo and Azoxy Compounds*, Vol. I, S. Patai, ed., New York, NY: Wiley, p. 259.
74. Ulrich, D. R. 1988. "Light-Induced Anisotropy in Azo-Dye-Colored Materials," *Mol. Cryst. Liq. Crys.*, 160:1–31.
75. Singer, K. D., J. E. Sohn and S. J. Lalama. 1986. "Second Harmonic Generation in Doped Polymer Films," *Appl. Phys. Lett.*, 49:248–250.
76. Fujishima, A., Z. F. Liu and K. Hashimoto. 1990. "Photoelectrochemical Information Storage Using an Azobenzene Derivative," *Nature*, 347:658.

3

AZO FUNCTIONAL POLYMERS–STRATEGIES FOR SYNTHESIS

3.1. Introduction

Development of novel synthetic procedures continues to occupy an increasingly important place in the repertoire of a polymer chemist as well. New synthetic methods and reactions characterized by molecular weight control and stereospecificity for a variety of end-uses are being continuously developed. Polymers with functional groups have become of major importance in the advancement of polymer science. The macromolecules are often tailor-made to perform specific functions and are sought for these properties rather than for the mechanical or bulk properties. These dual requirements offer stimulating challenge to a polymer chemist. Often, the best outcome will be a trade off between the demands of the functional performance and processability. The term polymer synthesis describes the process of creating a specific macromolecular architecture from an assembly of monomeric units using one or more polymerization processes. The macromolecular framework may be designed to contain functional groups in the side chains, backbone or in the crosslinks; the polymer chain may be linear, branched or crosslinked, the monomer composition may be varied to give hydrophilic or hydrophobic materials.

Much of the synthetic methodology available on the preparation of azo polymers can be conveniently classified and discussed on the basis of (a) location of the azo functional group on the polymer chain and (b) on the thermal stability of the azo group i.e., aliphatic and aromatic nature.[1]

In most cases, the synthesis involves selection of a polymerizable derivative of a corresponding azo compound and its polymerization to yield required structures. We have employed the term "azomer" to describe any *azo* group containing mono*mer* to include a mono or difunctional derivative which can undergo either condensation or addition polymerization to become a part of a macromolecular structure. For example, 4-vinyl azobenzene or 4,4′diamino azobenzene will be referred to as an azomer in the discussion. We have attempted to present here the general strategies rather than the synthetic details and the treatment is only qualitative. Some references have been made only in those cases where the extrapolation of "azo effect" to polymer matrix might result in interesting classes of polymers.

Two approaches exist for the preparation of functional polymers, namely the polymerization or copolymerization of monomers which carry the desired azo functionality (azomers) and the chemical modification of pre-formed polymer. The first approach is the more direct approach and many functional linear polymers have been prepared by either condensation type polymerization or by any of the additional polymerization procedures such as radical, cationic, or anionic polymerizations. The interest in this approach stems from the possibility of predicting the relative location of the functional group inside a polymer chain i.e., sequence distribution, when the reactivity ratios of the comonomers are known.

An alternative to direct copolymerization for preparing functionalized polymers is the chemical modification of preformed polymers. This method is particularly attractive for the preparation of crosslinked polymers because one can start with commercially available microporous or macroporous polymer beads of good physical size and form, with a known percentage of crosslinking and porosity, etc. Chemical modification is, however, not without its problems. Since few reactions are totally free from side reactions, one cannot afford to carry out chemical modification using reaction sequences with more than two or three steps. Both these strategies have been employed for the synthesis of azo functional polymers.

3.2. Aliphatic Azo Polymers

3.2.1. Azo Groups in the Main Chain

As early as 1951, the patent literature reported polymers containing azo groups.[1,2] Much of the interest in these aliphatic azo polymers stems from the need to design sequential and polyfunctional initiators which are thermally labile. From the chemical point of view, these compounds are characterized by the presence of at least two labile functional groups—either different or of the same type, which are in most cases, azo or peroxy groups. A poly functional initiator will be a sequential initiator if it can be decomposed in stepwise fashion. This is possible if the functional groups, generating free radicals are dissimilar and possess different thermal or photochemical stability. This property is essential for obtaining

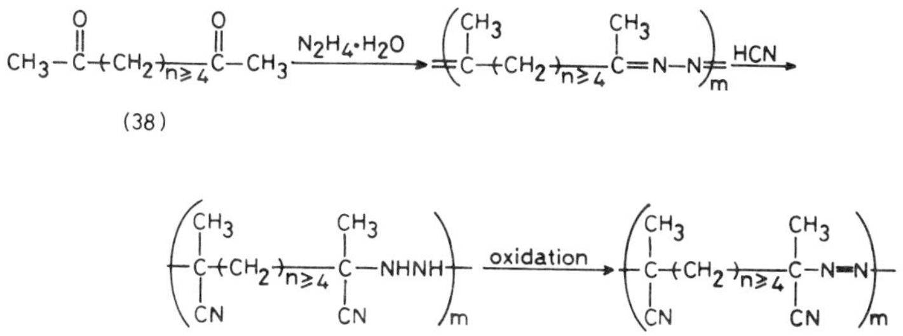

$$CH_3-\overset{O}{\overset{\|}{C}}-(CH_2)_{n\geq 4}-\overset{O}{\overset{\|}{C}}-CH_3 \xrightarrow{N_2H_4\cdot H_2O} \left(=\overset{CH_3}{\overset{|}{C}}-(CH_2)_{n\geq 4}-\overset{CH_3}{\overset{|}{C}}=N-N=\right)_m \xrightarrow{HCN}$$

(38)

$$\left(-\underset{CN}{\underset{|}{\overset{CH_3}{\overset{|}{C}}}}-(CH_2)_{n\geq 4}-\underset{CN}{\underset{|}{\overset{CH_3}{\overset{|}{C}}}}-NHNH-\right)_m \xrightarrow{\text{oxidation}} \left(-\underset{CN}{\underset{|}{\overset{CH_3}{\overset{|}{C}}}}-(CH_2)_{n\geq 4}-\underset{CN}{\underset{|}{\overset{CH_3}{\overset{|}{C}}}}-N=N-\right)_m$$

(39) (40)

FIGURE 3.1 Oxidation of polyhydrazines.

block copolymers, by a two-stage polymerization process. The general synthetic methods for the preparation of these polymeric initiators are discussed in this section while the thermal and photochemical decomposition of these polymers are discussed in subsequent chapters.

A. Modification of Preformed Polymer

a. Oxidation: The synthesis involves the conversion of aliphatic diketones having at least four carbon atoms between the carbonyl functions (38) to polyazines which are converted to polyhydrazonitriles(39) by addition of HCN, the latter are transformed into polymeric azo compounds (40) by oxidation.[2]

b. Nitroso Method: A completely different synthetic route involving nitroso polyamides yields polymers of the general structure (42), shown in Figure 3.2. Polyamides(41) containing N-nitroso groups should be handled carefully because they are photo- and thermally sensitive.

B. Copolymerization of Azomers

a. Condensation Polymerization: An alternative synthesis is to polymerize reactive AIBN analogs with suitable comonomers such as diamines, diols etc. According to MacLeay et al.[3,4] polyazo compounds are

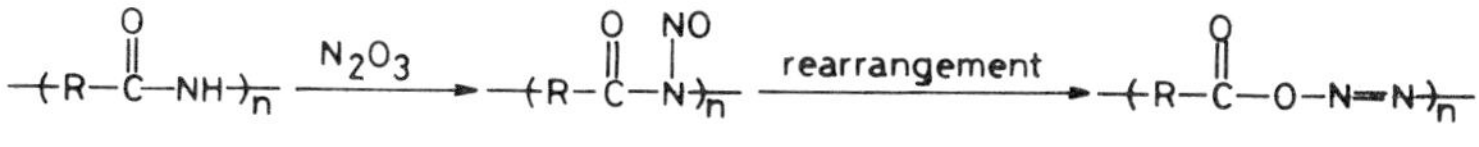

$$-\!\left(R-\overset{O}{\overset{\|}{C}}-NH\right)_n \xrightarrow{N_2O_3} -\!\left(R-\overset{O}{\overset{\|}{C}}-\overset{NO}{\overset{|}{N}}\right)_n \xrightarrow{\text{rearrangement}} -\!\left(R-\overset{O}{\overset{\|}{C}}-O-N=N\right)_n$$

(41) (42)

FIGURE 3.2 Rearrangement of N-nitroso polymers.

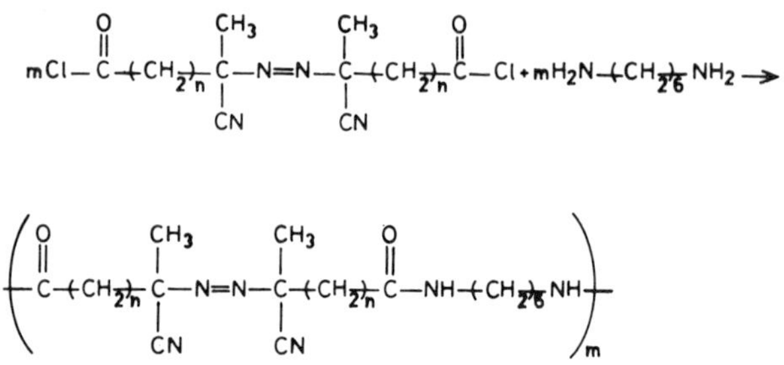

(43)

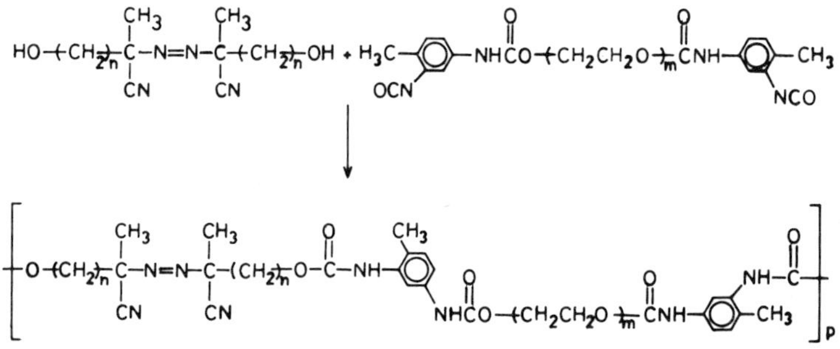

(44)

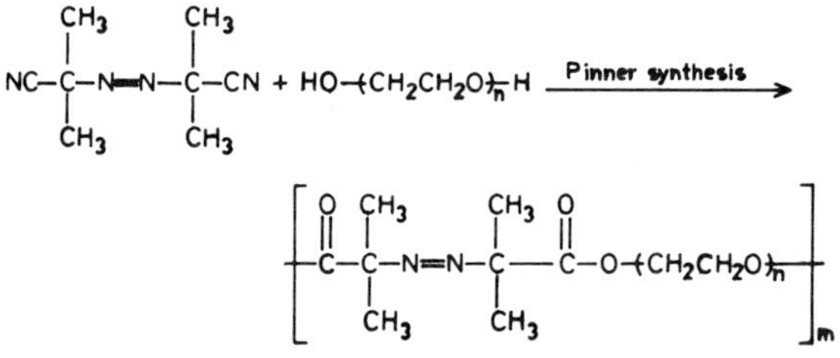

Pinner synthesis

(45)

FIGURE 3.3 Azo polymers by condensation polymerization and by Pinner synthesis.

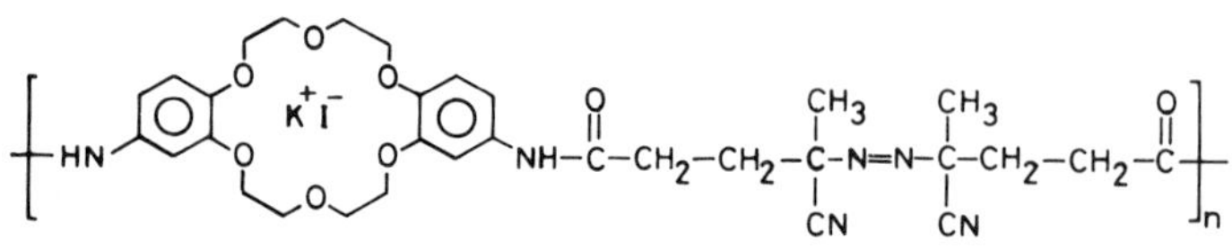

(46)

FIGURE 3.4 Aliphatic azo polymers with crown ether units.

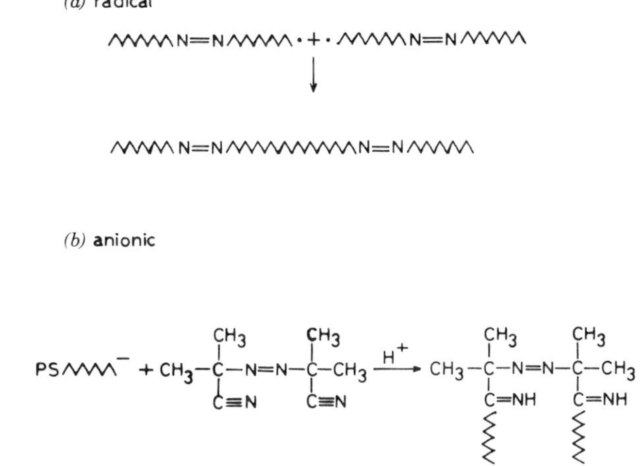

FIGURE 3.5 Aliphatic azo polymers with radical and anionic techniques.

obtained by a coupling between two or more monoazo derivatives. In order to effect this, one of the components needs to be an acylating function, the other one a reactive group (alcohol, thiol, amino) that allows the fragment bonding by formation of esters, thioesters, carbonates, thiocarbonates, amides(43), urethanes(44) etc. This copolymerization is an extremely versatile method.[5-10] A fine example of the versatility of the copolymerization is demonstrated by Pinner synthesis[11] which has been utilized recently in a study of the synthesis of polyesters(45).[12-17] Conversion of nitrile groups into ester groups allows the introduction of additional OH groups if diols in excess are used as shown below.

A number of interesting variations of the condensation polymerization are reported. Yagci et al.[18] reported the synthesis of a macroazoinitiator with crown ether units (46) by solution condensation reaction. Since, the complexation of the polymer supported crown ethers has great importance being the first fundamental process in phase transfer reactions.

b. Free-Radical Method: In addition to condensation techniques, azo polymers can be obtained via radical[19] and anionic[20,21] (Figure 3.5) polymerization techniques. From radical initiators, containing stable azo groups and less stable peroxy groups, azo copolymers can be produced. Polymers containing more than one azo group per chain can also be formed during the above reaction by radical recombination.

c. Cationic Polymerization: Ionic addition polymerization has also been described for the synthesis of polymers having the azo function in the main chain.[22,23] Besides condensation type reactions, low molecular weight

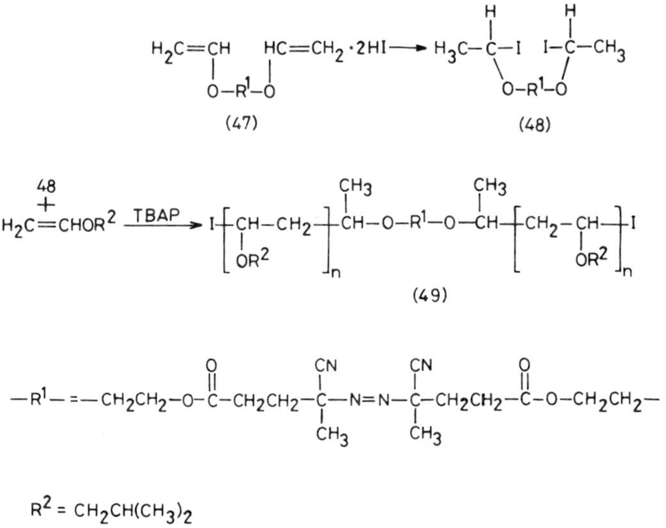

$$H_2C{=}CH \quad HC{=}CH_2 \cdot 2HI \longrightarrow H_3C{-}CH(I){-}O{-}R^1{-}O{-}CH(I){-}CH_3$$

(47) (48)

$$48 + H_2C{=}CHOR^2 \xrightarrow{TBAP} I{-}[CH(OR^2){-}CH_2]_n{-}CH(CH_3){-}O{-}R^1{-}O{-}CH(CH_3){-}[CH_2{-}CH(OR^2)]_n{-}I$$

(49)

$$-R^1- = -CH_2CH_2{-}O{-}C(=O){-}CH_2CH_2{-}C(CN)(CH_3){-}N{=}N{-}C(CN)(CH_3){-}CH_2CH_2{-}C(=O){-}O{-}CH_2CH_2-$$

$R^2 = CH_2CH(CH_3)_2$

FIGURE 3.6 Polymeric azo initiators via cationic polymerizations.

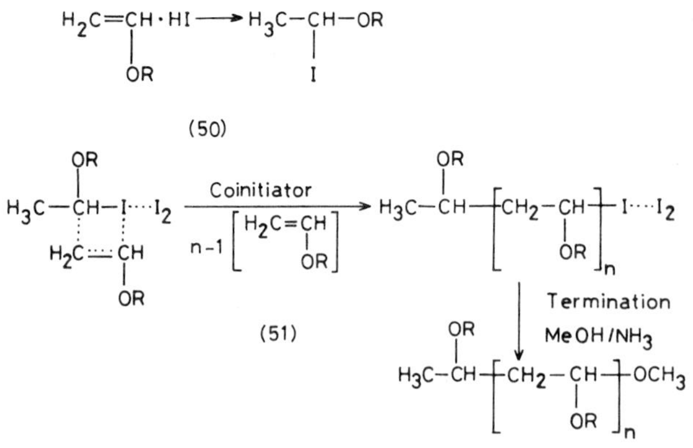

$$H_2C{=}CH(OR) \cdot HI \longrightarrow H_3C{-}CH(I){-}OR$$

(50)

$$H_3C{-}CH(OR){-}I\cdots I_2 \;\; (H_2C{=}CH{-}OR) \xrightarrow[n-1\,[H_2C=CH(OR)]]{\text{Coinitiator}} H_3C{-}CH(OR){-}[CH_2{-}CH(OR)]_n{-}I\cdots I_2$$

(51)

Termination MeOH/NH$_3$

$$\downarrow$$

$$H_3C{-}CH(OR){-}[CH_2{-}CH(OR)]_n{-}OCH_3$$

FIGURE 3.7 Mechanism of living cationic polymerization.

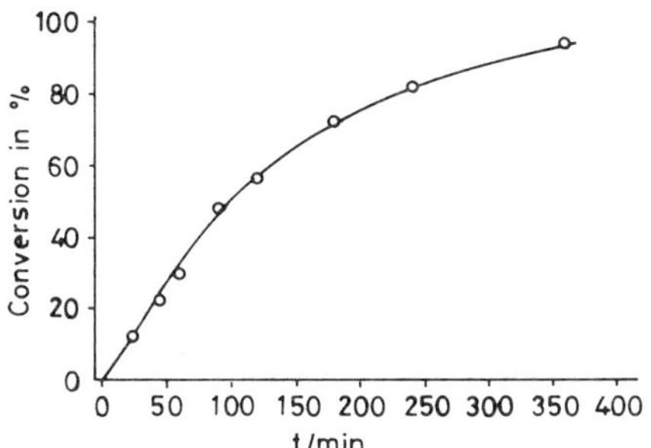

FIGURE 3.8 Conversion of isobutyl vinyl ether in the reactions with polymeric initiator (49) as a function of time in CH_2Cl_2 at 40°C.

azo compounds have been used in ionic polymerizations as initiators,[18] or terminating agents without loss of the azo function. A promising pathway is opened up by the application of living cationic polymerization to a special 1,3 divinyl ether containing a central thermosensitive azo function. The reaction sequence is shown in Figure 3.6. 2,2′ azobis [2-methyl-4-(2 vinyl oxy ethoxycarbonyl)butyronitrile](47) was considered to be a key compound for the synthesis of a macro azo initiator(49) using the living polymerization. Both azo compounds have asymmetric carbon atoms in their molecules. Due to the two vinyloxy groups, it was not surprising that crosslinking was observed during the copolymerization. Recently, Higashimura et al.[24,25] have found that cationic polymerization of vinyl ethers with HI/I_2 shows the characteristic of a living system: control of the molar masses by $[M]:[I]$, smaller molar mass distribution, and linear growth of the molar masses with conversion of the end groups by using suitable terminating agents or initiating molecules.

In the first step a calculated amount of dry HI was added to the vinyl ether, yielding the addition product 1-alkoxy-1-iodoethane. The carbon-iodine bond is then activated by a coinitiator, such as I_2, ZnI_2, allowing the unreacted monomer to insert into this bond.

Nuyken et al.[26,27] have modified this technique and prepared azo polymers using divinyl ether(50), adding sequentially HI, coinitiator and isobutyl vinyl ether(51) as monomer to it. The polymer chain grows into two directions as shown below. The time vs. conversion curve for the polymerization reaction at −40°C of (47) is shown in Figure 3.8. Mechanistic details of this modified living cationic polymerization are published elsewhere.[26]

d. End-Capping Technique: Another method of producing polymers containing one or more main chain azo group depending on the functionality of the growing chains is to terminate an ionic polymerization by the addition of AIBN[20] (Figure 3.5).

3.2.2. Azo Groups in Side Chains

A. Modification of Preformed Polymer

This method is not very successful if several steps are involved. The problem associated with the polymer modification route is demonstrated by the difficulty of introducing acetyl amino groups into styrene copolymers(52). In an attempt to reduce the number of steps 4-acetyl amino styrene(55) was used for the copolymerization.[29] The resulting copolymer can be converted to an azo polymer only in two steps viz., nitrosation followed by rearrangement. However, (54) is thermally unstable and cannot be isolated. Another example of these difficulties is the reaction between hydroxy containing polymers. Only crosslinked products are obtained, because azo functions react intermolecularly. For other polymer modifications, the reader may refer to patent literature.[30]

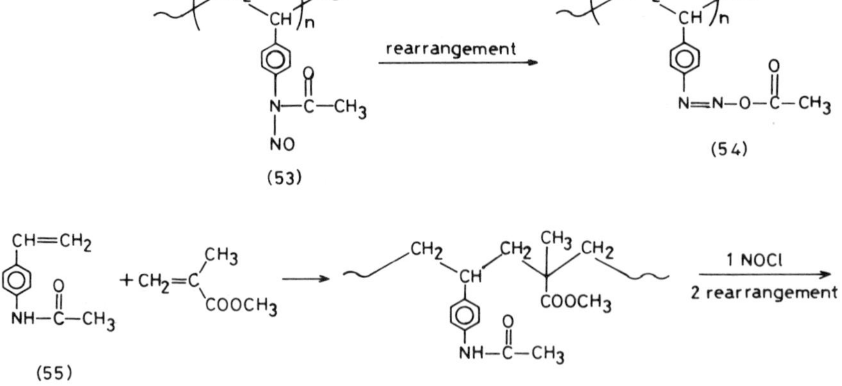

FIGURE 3.9 Aliphatic azo side chains by N-nitroso rearrangement.

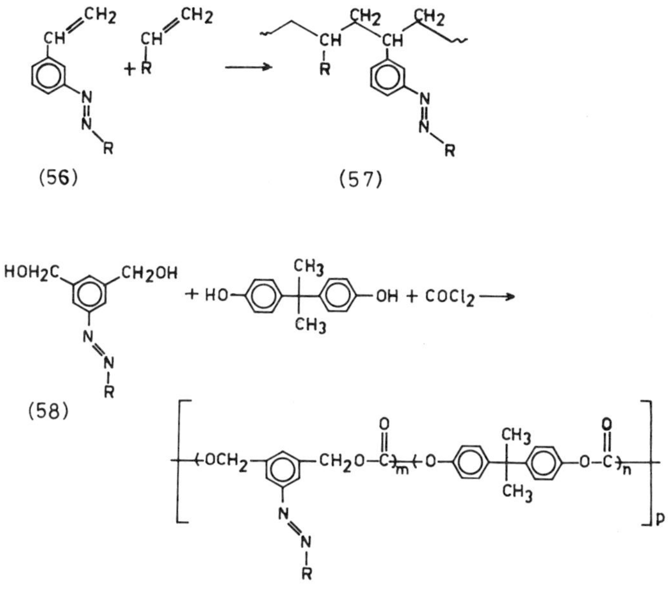

(56) (57)

(58)

(59)

FIGURE 3.10 Aliphatic azo side chains by additions and condensation polymerization methods.

B. Polymerization of Azomer

Linear azopolymers have been obtained via copolymerization of asymmetric azo monomers.[31-35] A wide variety of other structures (56,58) have been described in a patent.[30] Monomers shown in Figure 3.10 can be copolymerized with more common monomers such as styrene, MMA, etc. to give polymers with pendant azo groups. The number of azo groups in the polymer is controlled by the composition of the monomer mixture and the copolymerization conditions. An alternate procedure is demonstrated by the telomerization of styrene, butadiene and azo monomer systems.[36] In this method, the azo content of the resulting polymer is controlled by the composition of the monomer mixture.

3.3. Aromatic Azo Polymers

Many different synthetic routes are available for the production of polymers containing stable azo functions in the backbone. Most widely used methods are briefly mentioned here.

3.3.1. Azo Groups in the Main Chain

A. Polycondensation

Azo polyamides and polyesters can be synthesized via interfacial or solution polymerization techniques.[37-43] Using the basic reaction shown, a series of polyamides(**60**), polyesters(**61**), polyurethanes, and polyethers(**62**) have been synthesized by substitution of the azomer with other difunctional azo free units.

A little more specialized polycondensation reaction is an attempt to produce highly conjugated polymer,[44] by the reaction of 4,4′-diaminoazobenzene with terephthaldehyde to give a condensation product (**63**) with $n = 4$–6. In addition, a polymer(**64**) was produced by the reaction of 4,4′azo dibenzaldehyde with a bifunctional Wittig salt by the same workers.[44]

B. Polymerization through Azo Group Formation

a. Polydiazotization: Low temperatures are necessary for the formation of polymeric azo compounds by the coupling reaction of diazonium salts.[45-47] During the conversion of bis(benzidine) diazonium salts(**65**) into polyphenylenes, some $-N{=}N-$ units are also formed (**66**). The number of azo groups per chain can be controlled by changing the reaction

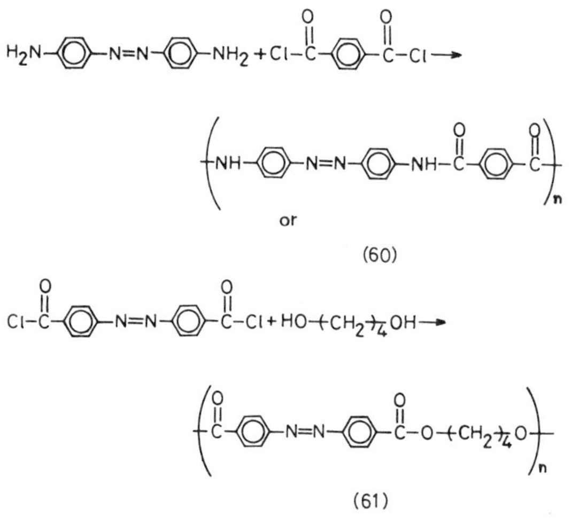

(60)

(61)

FIGURE 3.11 Aromatic azo polymers by polycondensation methods.

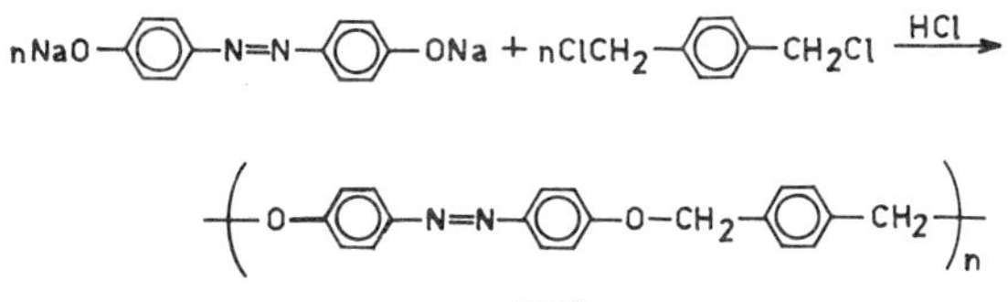

(62)

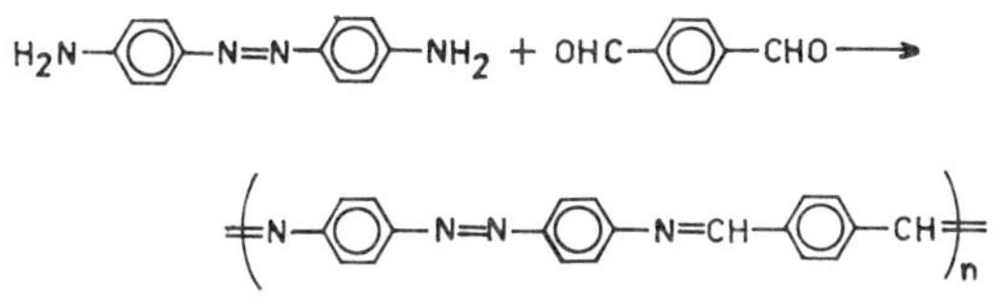

(63)

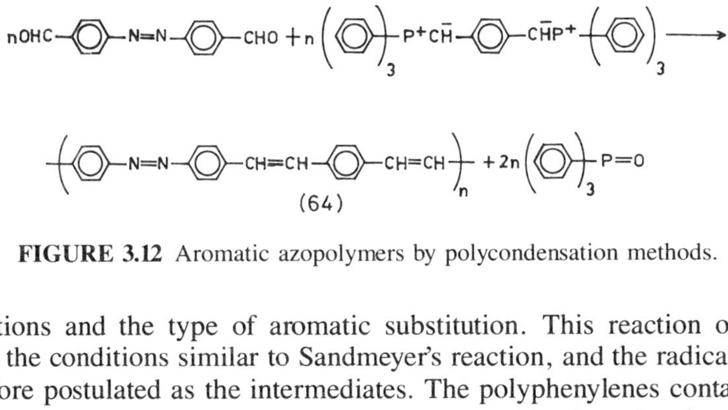

(64)

FIGURE 3.12 Aromatic azopolymers by polycondensation methods.

conditions and the type of aromatic substitution. This reaction occurs under the conditions similar to Sandmeyer's reaction, and the radicals are therefore postulated as the intermediates. The polyphenylenes containing azo functionalities can then form by radical recombination. Another strategy is to use photochemical reaction of bis azides to produce azoaromatic linkages(67).

b. Poly Oxidative Coupling: The oxidative coupling of organic molecules is rapidly emerging as a new and general field of polymer chemistry. In 1955, Mogiliyanski et al.[48] described a novel, catalyzed oxidative coupling reaction, the formation of azobenzene and azobenzene derivatives(69) from primary aromatic amines(68) such as aniline by oxygenation in pyridine in the presence of cuprous chloride (Figure 3.14). Since 1964 Kotlyarevski et al.[49] have reported on the formation of aromatic azo polymers by the above technique. Since then, the catalyzed oxidative coupling of aromatic diamines has been intensively used for the synthesis of polymers containing stable azo groups.[50]

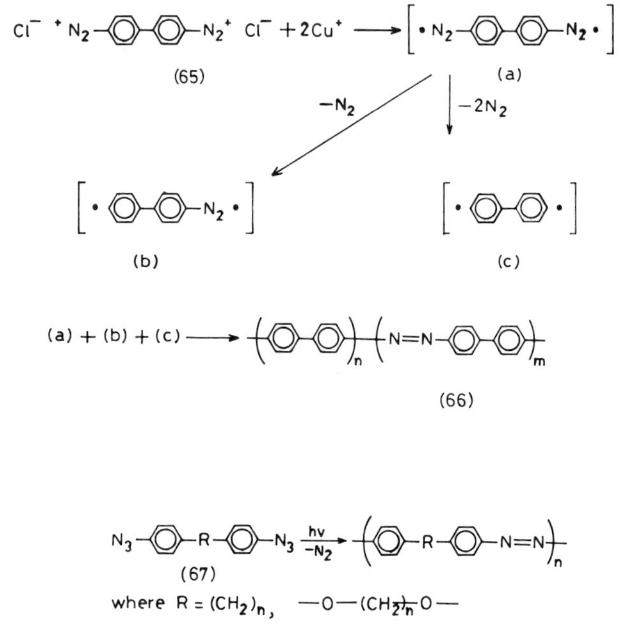

$Cl^- \; {}^+N_2$—⌬—⌬—$N_2^+ \; Cl^- + 2Cu^+ \longrightarrow [\cdot N_2$—⌬—⌬—$N_2 \cdot]$

(65) (a)

$-N_2$; $-2N_2$

$[\cdot$⌬—⌬—$N_2 \cdot]$ (b)

$[\cdot$⌬—⌬$\cdot]$ (c)

(a) + (b) + (c) ⟶ $\{$⌬—⌬$\}_n \{N{=}N$—⌬—⌬$\}_m$

(66)

N_3—⌬—R—⌬—$N_3 \xrightarrow[-N_2]{h\nu} \{$⌬—R—⌬—$N{=}N\}_n$

(67)

where $R = (CH_2)_n$, —O—$(CH_2)_n$—O—

FIGURE 3.13 Aromatic azo polymers by polydiazotization methods.

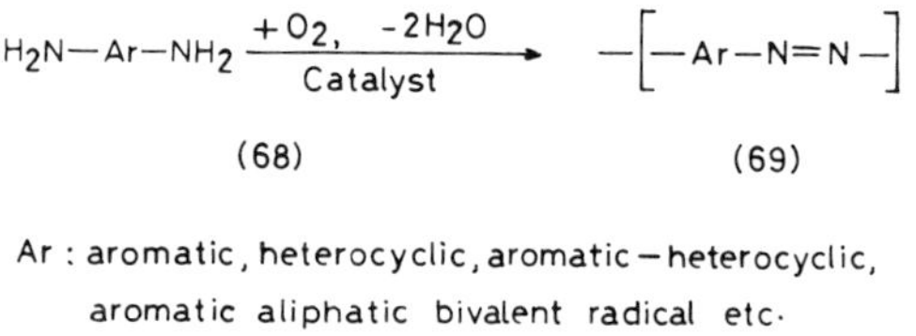

$$H_2N—Ar—NH_2 \xrightarrow[\text{Catalyst}]{+O_2,\; -2H_2O} —[—Ar—N{=}N—]$$

(68) (69)

Ar : aromatic, heterocyclic, aromatic – heterocyclic, aromatic aliphatic bivalent radical etc.

FIGURE 3.14 Oxidative coupling of aromatic diamines.

The reactions leading to the formation of these polymers have one feature in common, although they otherwise differ greatly in mechanism;[51-53] the crucial step in the reaction sequence is a one electron transfer from the monomer to a transition metal ion serving as an electron acceptor. In addition to being an electron acceptor the transition metal ion is also probably involved in the coupling reaction by complexation of radical-like intermediates produced.[53]

3.3.2. Aromatic Azo Groups as Side Chains

A. Modification of Preformed Polymer

Like in earlier structural types, the incorporation of azo aromatic moiety can be achieved by the modification of prepolymer or by the polymerization of azomer. Several classes of polymers have been designed in the area of polymeric dyes, liquid crystalline polymers and for developing prodrug systems. The most common strategy has been to employ the diazotization reaction as shown in Figure 3.15.

B. Polymerization of Azomers

Several monomeric (71) systems where the functional group is separated from the polymer chain by a spacer group have been reported as shown in Figure 3.16.[54,55] Very few examples of other azoheterocyclic systems with the formation of double bonds and their polymerization are reported in the literature.

3.3.3. Azo Groups in the Crosslinks

Recently, there are reports[56,57] of azo crosslinked polymers made by the incorporation of azo groups in the crosslinks by a radical polymerization technique. The synthesis of divinyl azobenzene(72) was carried out as shown in Figure 3.17.

Likewise, other crosslinking agents with spacer groups between the double bond and the azo moiety were prepared and polymerized to get photoresponsive polymers[57] and biodegradable polymers.[58,59]

3.4. Conclusion

This brief review has attempted to bring together the general principles of polymer design to incorporate azo functionality. It is clear that the

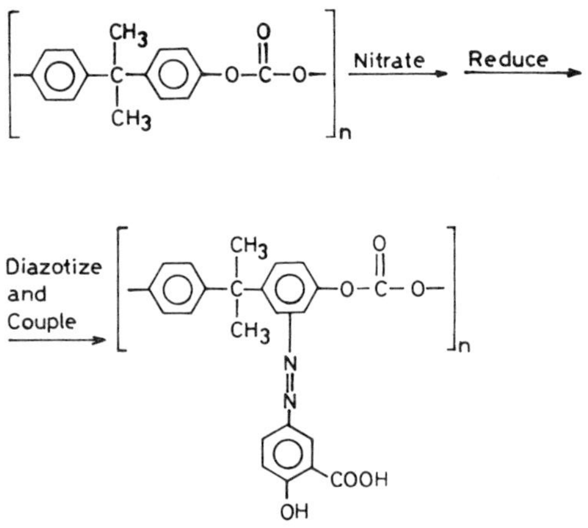

(70)

FIGURE 3.15 General scheme to introduce azoaromatic side chains.

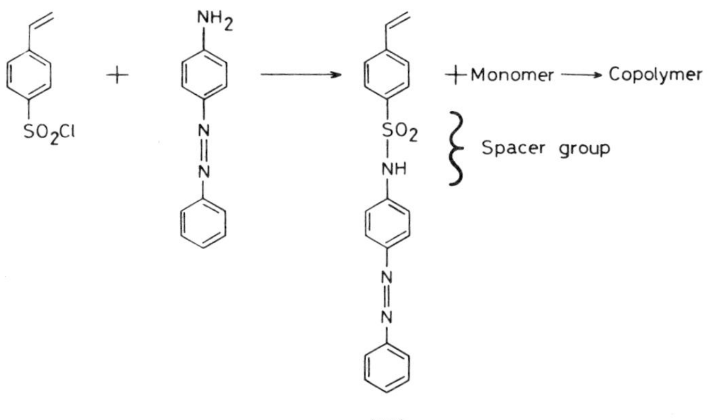

(71)

FIGURE 3.16 Azomers with spacer groups and their polymerization.

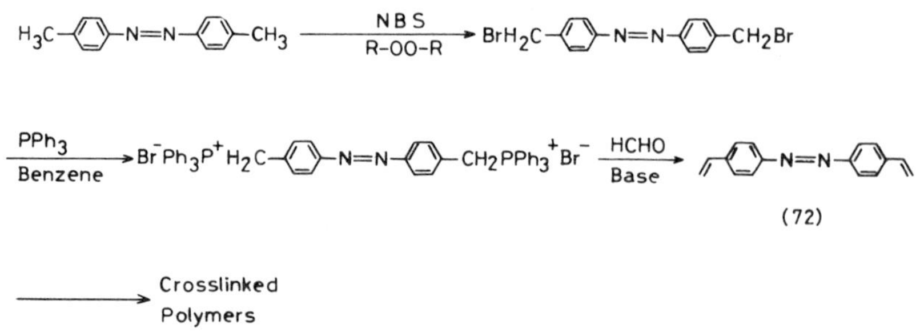

(72)

FIGURE 3.17 Synthesis of an azoaromatic crosslinker: 4,4′ divinyl azobenzene.

methods like condensation, diazotization, oxidative coupling etc., which have been used to prepare low-molecular weight azo compounds have been extended to prepare azo polymers. Reasoning by analogy is fundamental to organic synthesis, but there are quite definite contraints. In most polymer synthesis, it is usually necessary to construct polymer chains from smaller units. Although in principle there are several ways in which this can be done, in practice it is much more restricted by considerations of overall yield and conversions. The macromolecular assembly is usually accomplished by preparing azo functional monomers that constitute building blocks of the parent polymer and joining them together. The result is a more convergent, less linear synthesis. The need then is for a battery of methods for joining these pieces together. In view of the differing thermal stabilities of the aliphatic and aromatic azo moieties, the importance and relevance of these methods have been different. The preponderance of methods deal with azobenzene units and very little, if any, other azoaromatic units have been incorporated in the polymer chain. Similarly AIBN units dominate the aliphatic scenario. I believe that the scope is vast for a synthetic chemist to develop a vast array of functionalized aliphatic and aromatic azomers for a number of specific end-uses.

3.5. References

1. Nuyken, N. 1982. In *Encyclopedia of Polymer Sci. & Engineering, 2nd Ed.*, H. F. Mark, ed., New York, NY: Wiley & Sons, 2:158.
2. Hill, J. W. U.S. Pat. 2,556,876, 1951.
3. Shepparad, C. S. and R. E. McLeay. U.S. Pat. 3,987,024, 1976.
4. Shepparad, C. S. and R. E. McLeay. U.S. Pat. 4,075,286, 1978.
5. Heitz, W., H. G. Stahl and R. Dicke. Ger. Offen. 3,005,889, 1981.
6. Nagai, S., Y. Hidaka, and A. Ueda. Jpn. Kokai. 74,87,895, 1974.
7. Smith, D. A. 1967. "The Thermal Decomposition of Azonitrile Polymers," *Makromol. Chem.*, 103:301.
8. Chandalia, K. B. and F. J. Preston. U.S. Pat. 4,094,868, 1978.
9. Laverty, J. J. and Z. G. Gardlund. 1974. "Poly(vinyl chloride) Block Copolymers by Macrobiradical Initiation," *Polym. Prepr., Amer. Chem. Soc. Div. Polym. Chem.*, 15:306.
10. George, M. H. and J. R. Ward. 1973. "Polymerization of Styrene by Poly-(Bisphenol A 4,4′-Azobis-4-cyanopentanoate)" *J. Polym. Sci. Polym. Chem. Ed.*, 11:2909.
11. March, J. 1977. *Advanced Organic Chemistry, 2nd Ed.*, New York, NY: McGraw-Hill, p. 813.
12. Furukawa, J., S. Takamori and S. Yamashita. 1967. "Preparation of Block Copolymers with a Macro Azonitrile as an Initiator," *Angew. Makromol. Chem.*, 1:92.

13. Walz, R. and W. Heitz. 1978. "Preparation of $(AB)_n$-Type Block Copolymers by the Use of Azopolyesters," *J. Polym. Sci. Polym. Chem. Ed.*, 16:1807.
14. Walz, R., B. Bomer and W. Heitz. 1977. "Monomeric and Polymeric Azoinitiators," *Makromol. Chem.*, 178:2527.
15. Heitz, W., M. Lattekamp, C. Oppenheimer and P. S. Anand. 1983. "Synthesis of Block Sequences by Radical Copolymerization," *ACS Symposium Series*, Washington DC: ACS., 212:337.
16. Dicke, H. R. and W. Heitz. 1981. "Synthesis and Characterization of Surface-active Azo-group Containing Polyacrylamides," *Makromol. Chem. Rapid. Comm.*, 2:83.
17. Anand, P. S., H. G. Stahl, W. Heitz, G. Weber and L. Bottenbruch. 1982. "Block Copolymers by Radical Polymerization," *Makromol. Chem.*, 183:1685.
18. Yagci, Y., U. Tunca and N. Bicak. 1986. "A New Macro-azo Initiator for the Synthesis of Polymers with Crown Ether Units," *J. Polym. Sci. Polym. Lett.*, 24:491.
19. Plirma, I. and B. Gunesin. 1977. "Kinetics of Butyl Acrylate Polymerization with PMMA Polymeric Initiator," *Polym. Prepr., Am. Chem. Soc. Div. Polym. Chem.*, 18:687.
20. Vinchon, Y., R. Reeb and G. Reiss. 1976. "Anionic Preparation of Azo Macromolecular Promoters–Application to the Synthesis of Block Copolymers," *Eur. Polym. J.*, 12:317.
21. Reiss, G. and R. Reeb. 1981. "Preparation of Polymeric Free Radical Initiators by Anionic Synthesis of Polymeric Azo Derivatives," *ACS Sympo. Ser., Washington DC*, 166:477.
22. Furukawa, J., S. Takamori and S. Yamashita. 1967. "Preparation of Block Copolymers with a Macroazonitrile Initiator," *Agnew. Makromol. Chem.*, 1:92.
23. Vollmert, B. and H. Bolte. 1960. "Preparation of Graft Copolymers by Azonitrile Groups," *Makromol. Chem.*, 36:17.
24. Miyamoto, M., M. Sawamoto and T. Higashimura. 1984. "Synthesis of Monodisperse Living Poly (vinyl ethers) and Block Copolymers by the Hydrogen Iodide/Iodine Initiating System," *Macromolecules*, 17:265.
25. Miyamoto, M., M. Sawamoyo and T. Higashimura. 1985. "Synthesis of Telechelic Living Poly (ethers)," *Macromolecules*, 18:123.
26. Nuyken, O., H. Kroner and S. Aechter. 1988. "Synthesis of Living Telechelic Poly(ethers) II," *Makromol. Chem. Rapid. Commun.*, 9:671.
27. Nuyken, O. and R. Weidner. 1986. "Graft and Block Copolymers via Azo Polymeric Initiators," *Adv. Polym. Sci.*, 73/74:147.
28. Craubner, H. 1982. "Macromolecular N-nitroso Acylamines as Photosensitizers for Polyreactions," *J. Polym. Sci. Polym. Chem. Ed.*, 20:1935.
29. Hahn, W. and A. Fischer. 1956. "Polyfunctional Peroxides as Initiators of Vinyl Polymerization," *Makromol. Chem.*, 21:77.
30. Sheppard, C. S. and R. E. MacLeay. U.S. Pat. 4,218,370, 1980.
31. Kerber, R., O. Nuyken and R. Steinhausen. 1976. "Azo Initiators 3. Effects of Solvents on the Thermal Degradation of p-Methoxy phenyl azo-2-methylpropanedinitrile," *Makromol. Chem.*, 177:1357.
32. Kerber, R., O. Nuyken and M. Dorn. 1978. "Azo Initiators 7. Copolymerization

of (3-vinyl phenyl azo) methyl Malonodinitrile with Styrene," *Makromol. Chem.*, 179:1803.

33. Kerber, R., J. Gereem and O. Nuyken. 1979. "Azo Initiators 9. Synthesis and Grafting of Polycarbonates Containing Azo Groups," *Makromol. Chem.*, 180:609.
34. Nuyken, O., R. Rengel and R. Kerber. 1980. "Copolymerization of 4(1,1 dicyanoethyl azo) benzyl methacrylate with Styrene, Methacrylonitrile, and Methyl methacrylate and Subsequent Grafting of the Azo Copolymers," *Makromol. Chem.*, 181:1565.
35. Nuyken, O., M. Hofinger and R. Kerber. 1979. "Copolymers from 1-Phenyl-2-(3-vinyl phenyl thio)-diazene and Styrene and Grafting with Acrylonitrile," *Polym. Bull.*, 1:679.
36. Nuyken, O. and R. Weidner. 1988. "Azo Polymers—Synthesis and Reactions," *Makromol. Chem.*, 189:1331.
37. Agolini, F. and F. P. Gay. 1970. "Synthesis and Properties of Azoaromatic Polymers," *Macromolecules*, 3:349.
38. Blair, H. S., H. I. Pague and E. Riordan. 1980. "Photoresponsive Effects in Azopolymers," *Polymer.*, 21:1195.
39. Balasubramanian, M., M. J. Nanjan and M. Santappa. 1979. "Synthesis of Polyamides Containing Azo Group in the Polymer Chain," *Makromol. Chem.*, 180:2517.
40. Balasubramanian, M., M. J. Nanjan and M. Santappa. 1981. "Synthesis of Polyamides Containing 4,4′-azodibenzamido Units," *Makromol. Chem.*, 182:853.
41. Kumar, G. S., P. DePra and D. C. Neckers. 1984. "Chelating Polymers Containing Photosensitive Functionalities—I," *Macromolecules*, 17:2463.
42. Kumar, G. S., P. DePra, K. Zhang and D. C. Neckers. 1984. "Chelating Polymers Containing Photosensitive Functionalities," *Macromolecules*, 17:1912.
43. Irie, M. and H. Tanaka. 1983. "Photoresponsive Polymers 5. Reversible Solubility Change of Polystyrene Having Azobenzene Pendant Groups," *Macromolecules*, 16:210.
44. Kossmehl, G. and R. Wallis. 1982. "Polymeric Azomethines with Azobenzene Units," *Makromol. Chem.*, 183:347.
45. Berlin, A. A., G. V. Belova and I. V. Gudvilovich. 1967. "Polymers with Conjugated Systems," *Vysokomol Soedin*, Ser A 9:2214.
46. Parini, V. P. and I. V. Gudvilovich. 1965. "Polymers Formed by Azocoupling Reactions," *Izv. Akad. Vauk SSR, Ser Khim.*, 370:1965.
47. Berlin, A. A., B. I. Liogonski and V. P. Perini. 1961. "Preparation and Properties of Some Aromatic Compounds," *Vysokomol. Soedin*, 2:689.
48. Terentev, A. P. and Ya. D. Mogiliyanski. 1955. "Catalytic Autoxidation of Primary Aromatic Amines in the Presence of Complex of Pyridine and Cuprous Ion," *Dokl. Akad. Nauk SSSR*, 103:91.
49. Kotlyarevskii, I. L., M. S. Shvartsberg, L. B. Fischer, A. S. Zanina, M. Brdaova and M. P. Terpugova. 1968. "Highly Unsaturated Polymers of Polyethynyl Polyarene and Polyazoarene Series," *J. Polym. Sci.*, Part C 16:3803.

50. Bach, H. C. and W. B. Black. 1967. "Aromatic Azopolymers Produced by Oxidative Coupling of Primary Aromatic Diamines," *J. Polym. Sci.*, Part-C 22:799.
51. Bach, H. C. 1970. "Highly Ordered Azoaromatic Polyamides through a Novel Dimerization by Oxidative Coupling," *Polym. Prepr.* ACS Div. Polym. Chem., 11(1):334.
52. Bach, H. C. U.S. Pat. 3,637,534, Jan. 25, 1972.
53. Bach, H. C. and W. B. Black. 1969. "Polymerization of Primary Aromatic Diamines to Azopolymers by Oxidative Coupling," *Adv. Chem. Ser.*, 91:679.
54. Lovrien, R., J. C. B. Waddington. 1964. "Photoresponsive Systems I: Photochromic Macromolecules," *J. Am. Chem. Soc.*, 86(12):2315.
55. Eisenbach, C. D. 1978. "Effect of Polymer Matrix on the Cis-Trans Isomerism of Azobenzene Residues in Bulk Polymers," *Makromol. Chem.*, 179:2489.
56. Kumar, G. S., C. Savariar, M. Saffran and D. C. Neckers. 1985. "Chelating Polymers Containing Photosensitive Functionalities–III," *Macromolecules*, 18:1525.
57. Gaonkar, S. R., G. S. Kumar and D. C. Neckers. 1990. Photochromism of Azobenzene-Containing Polymers IV: Effect of Spacer Groups," *Macromolecules*, 23:5146.
58. Saffran, M., G. S. Kumar, D. C. Neckers, J. Jena, R. H. Jones and J. B. Field. 1990. "Biodegradable Azopolymer Coating for Oral Delivery of Peptide Drugs," *Biochem. Soc. Trans.*, 18:752.
59. Saffran, M., G. S. Kumar, C. Savariar, J. C. Burnham, F. Williams and D. C. Neckers. 1986. "A New Approach to the Oral Administration of Insulin and Other Peptide Drugs," *Science*, 233:1081.

4

THERMAL CHEMISTRY OF AZO FUNCTIONAL POLYMERS

4.1. Introduction

Considerable activity in polymer research focusses on materials that feature high strength, thermal stability and other structural properties under extreme environmental conditions. On the other hand, polymers are frequently used as not only structural components but also as selectively degradable carriers for other chemicals. In the latter applications, it is desirable to have matrix materials that retain their physical properties under one set of environmental conditions, but can be deliberately degraded in a controlled fashion by change of conditions. Heat-resistant polymers are used in many applications, such as insulators, for microelectronic components, sealants for fuel tanks in high speed aircraft, coatings on cookware, binders in brake systems and as structural components in space vehicles[1,2]. Thermolabile polymers are used as matrices for pigments (in paints and coatings) and for high energy compounds[3] (in explosives and propellents). The development of thermolabile and thermostable polymers increased enormously, as normally happens in the polymer industry, because of technical necessity of meeting these conflicting material requirements especially in the aerospace industry and all the requirements that went with its development.

As discussed in detail in Chapter 2, the azo functional group is unique in the sense that the substitution on the azo moiety decides the thermal behavior of the resulting compound.

Aliphatic azo compounds being thermally unstable decompose to give free radicals while aromatic azo compounds are extremely resistant to temperature due to resonance stabilization. The chemical factors which influence thermal behavior of these polymers include primary bond strength, secondary or van der Waals bonding forces, hydrogen bonding, resonance stabilization, mechanism of bond cleavage, rigid intrachain structure, crystallinity, etc.[4,5] It is the dichotomous nature of the azo functionality which forms the basis for the thermal chemistry of azo functional polymers.

4.2. Thermolabile Polymers: A Radical View

The most important property of the aliphatic azo moiety is its thermal instability and it is this property which has been made use of in designing novel classes of free-radical initiators. Azo compounds as initiators for free radical polymerizations were introduced by Lewis and Matheson in 1949.[6] However, the study of high molecular weight azo initiators containing more than one azo group per molecule has been relatively neglected, apart from a limited study of such molecules in solution, in the absence of added vinyl monomer. This is surprising since a study of vinyl polymerization initiated by a polyfunctional linear azo molecule, could *a priori*, give information about the relative reactivity of azo groups in different positions along the initiator chain and about the importance of "cage effects" and primary radical termination. Even in a simple polymerization, polyfunctional initiators provide the advantage of forming polymers of high molecular weights with increased initiator concentration and high rates of initiation.[7-14]

Secondly, sequential bi- and polyfunctional azo initiators have aroused great interest, owing to their special property of stepwise generation of free radicals, and implicitly, to the possibility they offer for being used in direct processes for block copolymer production.[15] From the chemical view point, these new initiators are characterized by the presence of at least two labile functionalities of different thermal stability, which like traditional initiators, are in most cases, azo and peroxy groups.

Polymers containing thermolabile azo functions are of three principal types. The azo functions can be incorporated into (a) the backbone, (b) the side chains or (c) they form the end groups.

The thermolysis of the azo functions of polymers of types (a) and (c) in the presence of a monomer leads to the formation of block copolymers, where as from polymers of type (b), a graft copolymer will result. How-

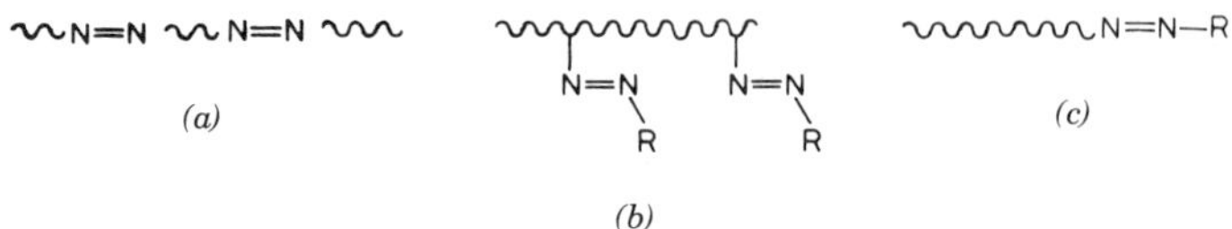

FIGURE 4.1 Polymers with thermolabile azo groups.

Table 4.1 Rate Constants (K_d) of Thermolysis and Initiator Efficiency of Monomeric and Polymeric Azo Compounds.

Temp (°C)	Solvent	BPA[a]	PEG 300/ AIBN[b]	ACV[c]	AIBN[d]
60	DMF	2.08		1.33	1.04
85	o-xylene	—	21	—	27.5
95	o-xylene	—	73	—	96

[a]BPA = poly(bisphenol A-4,4′azobis-4-cyanopentanoate); f = 0.27.
[b]PEG 300/AIBN polymer from polyethylene glycol and AIBN; f = 0.3.
[c]ACV = 4.4′-azobis-4-cyanopenanoic acid; f = 0.61.
[d]f = 0.57.

ever, it should be noted that the radical $R^{\bullet}$ from the thermolysis of the azo groups in polymers of types (b) and (c) will lead to homopolymer formation in addition to the desired graft or block copolymers.

Characteristics of azopolymers which determine the peformance are the average number of azo groups per chain, the rate constant of thermal decomposition and the initiator efficiency of the azo function (Table 4.1). Elemental analysis, UV spectroscopy, DSC and volumetric analysis, are employed to determine the azo content.[17-19] The thermolysis constant k_d is determined by UV and DSC.[18] Although the thermolysis constants of low molecular weight compounds and polymer analogs are essentially the same, the initiator efficiency of the polymeric azo compounds is generally lower.[20]

4.2.1. Type A: Main Chain Scission

A series of macroinitiators reported in the literature have the structure of polyazonitriles. In their synthesis, the basic intermediates are a bis azo alcohol and a diisocyanate (Chapter 2). On thermolysis, the macromolecular initiator generates free radicals capable of initiating polymerization. Many reports in the literature focus on the structural units which have been in traditional use. The repeating unit of azo bisisobutyronitrile dominates the aliphatic scenario possibly because of the ease of synthesis and availability of the starting material.

A. Thermal Decomposition: Kinetic and Mechanistic Studies

George and Ward[21] reported the polymerization behavior of styrene in DMF using a polymeric azo initiator, poly(bisphenol A 4,4′-azobis-4-cyanopentanoate)(73) (BPA). Polymerization rates were measured at

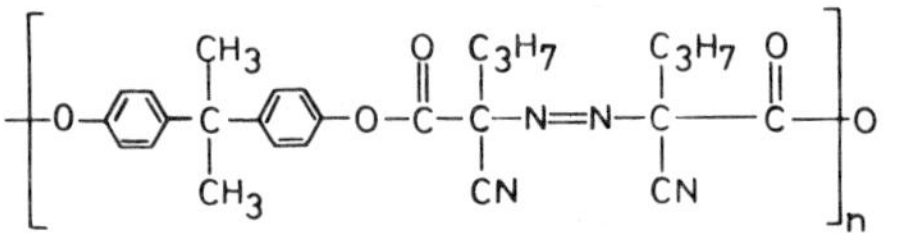

(73)

FIGURE 4.2

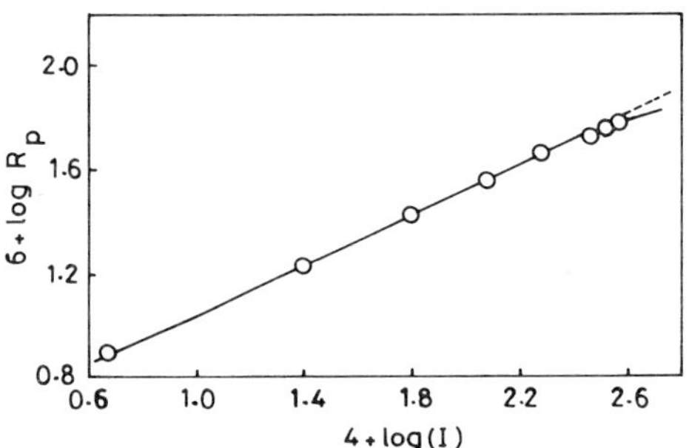

FIGURE 4.3 Plot of log R_p as a function of log $[I]$ for the polymerization of styrene in DMF at 60°C in the presence of BPA. $[M]_0 = 4.0$ mole/liter in all experiments.

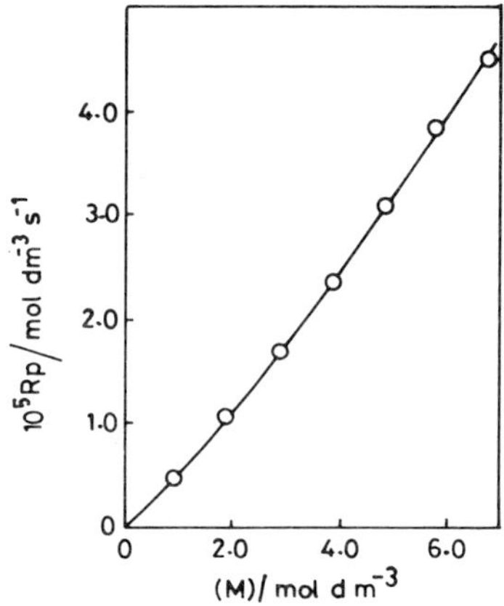

FIGURE 4.4 Plot of R_p as a function of $[M]$ for the polymerization of styrene in DMF at 60°C in the presence of BPA. $[I]_0 = 4.967 \times 10^{-3}$ moles/liter.

60°C gravimetrically or dilatometrically, and the molecular weights of isolated polystyrenes were determined viscometrically, both before and after hydrolysis. BPA has relatively low initiator efficiency of 0.28. The E_a and reaction velocity constant at 60°C for decomposition of BPA per azo group in DMF were spectroscopically found to be 105.9 cal/mole and 2.08×10^{-5} sec^{-1} respectively.

This system is especially interesting because comparative rate and molecular weight data were available for initiation of styrene polymerization in DMF by the monofunctional AIBN[22,23] and 4,4′azobis-4-cyanopentanoic acid[24] (ACV). A plot of log R_p against log $[I]$ at constant $[M]_0$ for BPA is shown in Figure 4.3. Where $[I]$ refers to the mean concentration of azo groups in the system. There is a gradient of 0.49 up to $[I] = 1.9 \times 10^{-2}$ mole/liter and above this concentration there was a slight downward trend. This behavior closely resembles that of ACV in DMF.[24] Likewise, a plot of R_p against $[M]$ at constant $[I]$ (Figure 4.4) shows that the order of the rate in monomer concentration is 1.16 when BPA is used, which is lower than 1.23 for AIBN[22] and 1.22 for ACV[24] and only slightly higher than the theoretical value of unity.

The kinetic study of the thermolysis of polyfunctional initiators is very important in testing the sequential character of the initiator, and implicitly in determining the thermal regime of the polymerization, starting from the decomposition kinetic parameters. The thermal decomposition kinetics have been followed in other macromolecular type initiators with azo groups inserted in the main chain. For a series of azo polyesters obtained from poly(ethylene glycol) (PEG) and AIBN, it has been proved that fragmentation takes place according to a first-order equation. From the Arrhenius plot of the decomposition rate constants, the following relation has been established.

$$k_d/s^{-1} = 1.99 \times 10^{-4} \exp(124.OKJ/RT)$$

The k_d values are presented in Table 4.2. Block copolymers are obtained from these initiators either when the AIBN modified to an oligomeric or

Table 4.2 Thermal Decomposition of the Polymer from PEG and AIBN.

Temperature (°C)	$K_d \cdot 10^4$ (sec^{-1})	$t_{1/2}$ (min)
85	2.106	54.8
88	3.500	33.0
95	7.312	15.8
105	19.950	5.8

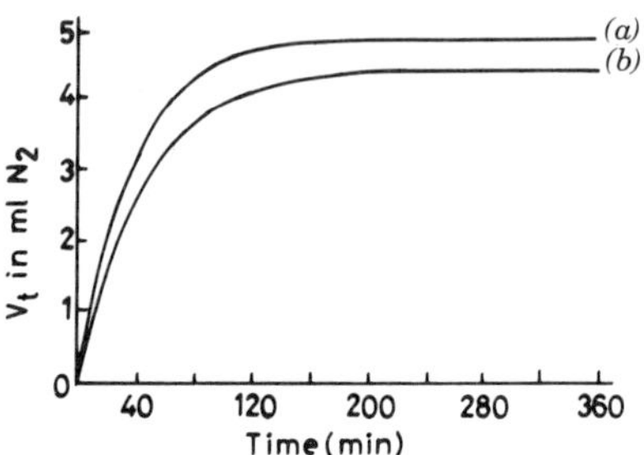

FIGURE 4.5 Time dependence of the decomposition of the polymer from (a) PEG 300 and AIBN and (b) of AIBN in O-xylene at 88°C.

polymeric diol is used or by a two-step polymerization similar to the procedure of Smets et al.[14]

According to MacLeay and Sheppard,[25-27] the successive preferential decomposition of the polyazo initiators is based on the thermal decomposition rate difference of the azo groups, at a constant temperature or at variable reaction times. For a series of di and polyazo derivatives, there have been established half-life times, useful in the evaluation of the thermal decomposition temperature of the component labile groups; they are quantitatively determined by traditional procedures such as GC, UV etc. The majority of these indices can be predicted approximately from the structure of the analogous monomers. For a series of azo initiators, the half-life times have been correlated with their chemical structure.[15]

Polymerization with polyfunctional initiators may be described by the conventional scheme. The considerable increase in the molecular weights with the polyfunctional initiators suggests a polymerization-recombination mechanism implying the following steps.

B. Cage Effect

In the homolytic decomposition of azonitrile compounds, the extent of radical combination is often in the range of 50–80%. This phenomenon is attributed to the "cage effect"[28] in which solvent molecules restrain the diffusion of radicals, so that combination becomes more probable. In his pioneering work, Smith[29] reported condensation type polyazo initiators(74) containing the unit (75) and studied their decomposition. In this way, the additional restraint of the polymer chain has been imposed on the diffusion of reactive radicals and differences in behavior have been observed, depending upon the nature of the connecting chain.

The evaluation of the cage effect has usually been made by isolation of

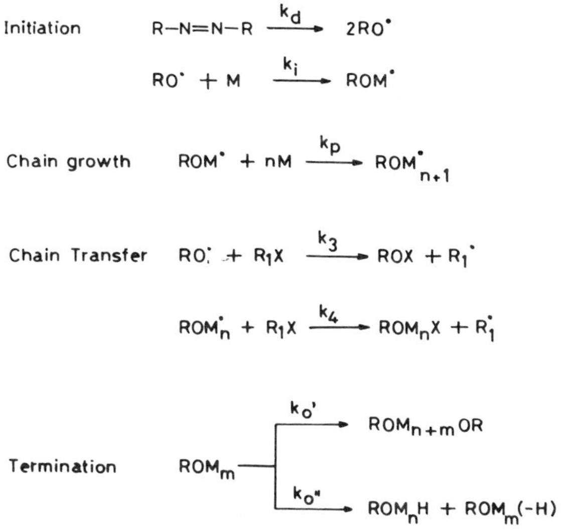

Initiation $R-N=N-R \xrightarrow{k_d} 2RO^{\bullet}$

$RO^{\bullet} + M \xrightarrow{k_i} ROM^{\bullet}$

Chain growth $ROM^{\bullet} + nM \xrightarrow{k_p} ROM^{\bullet}_{n+1}$

Chain Transfer $RO^{\bullet} + R_1X \xrightarrow{k_3} ROX + R_1^{\bullet}$

$ROM^{\bullet}_n + R_1X \xrightarrow{k_4} ROM_nX + R_1^{\bullet}$

Termination $ROM_m \xrightarrow{k_o'} ROM_{n+m}OR$

$ROM_m \xrightarrow{k_o''} ROM_nH + ROM_m(-H)$

FIGURE 4.6 Polymerization mechanism of polymeric initiators.

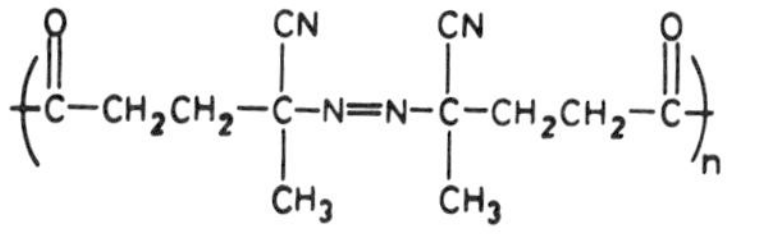

$$\left[-C(=O)-CH_2CH_2-C(CN)(CH_3)-N=N-C(CN)(CH_3)-CH_2CH_2-C(=O)- \right]_n$$

(74)

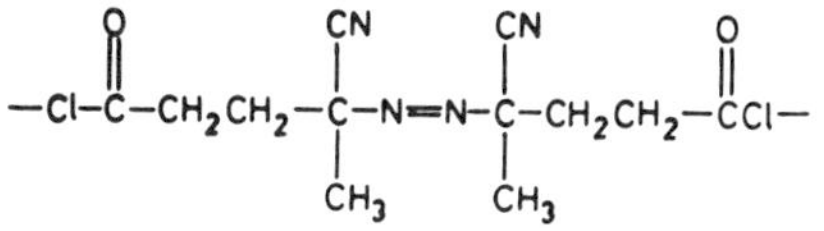

$$-Cl-C(=O)-CH_2CH_2-C(CN)(CH_3)-N=N-C(CN)(CH_3)-CH_2CH_2-C(=O)Cl-$$

(75)

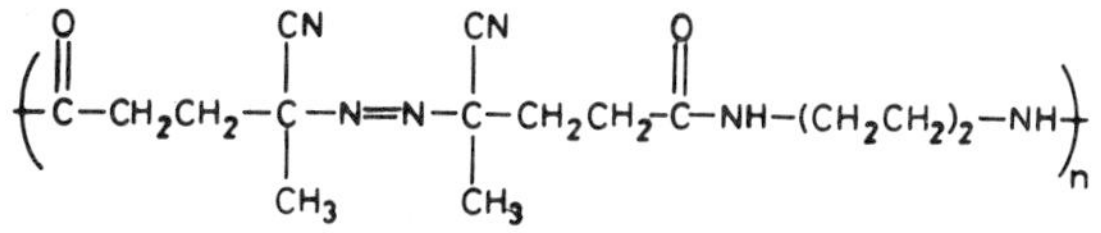

$$\left[-C(=O)-CH_2CH_2-C(CN)(CH_3)-N=N-C(CN)(CH_3)-CH_2CH_2-C(=O)-NH-(CH_2CH_2)_2-NH- \right]_n$$

(76)

$$\left[-C(=O)-CH_2CH_2-C(CN)(CH_3)-N=N-C(CN)(CH_3)-CH_2CH_2-C(=O)-NH-CH_2-C_6H_{10}-CH_2-NH- \right]_n$$

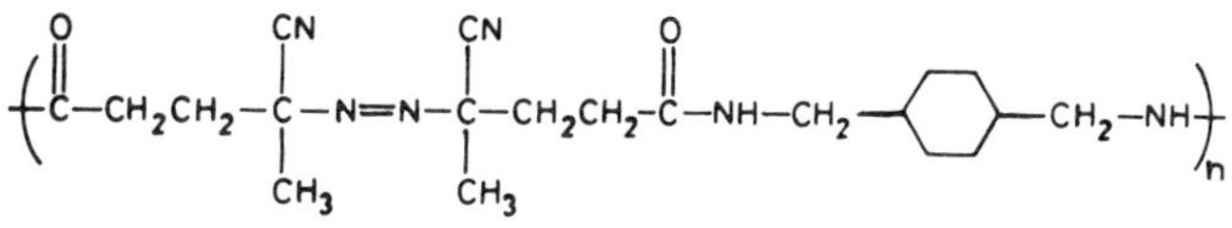

(77)

FIGURE 4.7

products or by entrapment of radicals by using efficient scavengers.[28,30] In this work, the inherent viscosity of the polymer was used as a sensitive test for the detection of very high degrees of radical combination. The synthesis of 4,4′-azo bis(4-cyanopentanoyl) chloride(75) was accomplished by the treatment of the corresponding acid with phosphorous pentachloride under very mild conditions. Representative condensation polymers were prepared by the mild interfacial method and subjected to thermal decomposition in solution. Polymers of the type (76) showed an extreme decrease in the viscosity, but those polymers containing rings in the chain (77) exhibited almost complete retention of viscosity. These results indicate that the diffusion of the radicals out of the solvent cage is strongly influenced by the rigidity of the chain and that intrachain units like cyclohexyl and diphenylmethyl appear to delay diffusion until combination takes place. In the case of polymer(71), the retention of viscosity is virtually quantitative; the combination reaction must therefore take place exclusively, since even a small percentage of scissions would give a substantial drop in viscosity. Smith also reported that the addition of a radical scavenger such as mercaptan did not affect the viscosity change significantly, since the alternative processes are combination and diffusion, not combination and capture.

C. Degradable Polymers

Kenley et al.[31] have considered thermally labile polymers that degrade simply upon heating. They have evaluated a difunctional azo compound azobis (3-hydroxy propyl diisobutylmethane) (BHPA)(78) as a monomer suitable for incorporation into polyurethanes.[32] They have demonstrated the feasibility of preparing thermally degradable polyurethanes that incorporate this repeating unit. Their work establishes important fundamental principles for designing thermally labile polymers of controlled lifetime. Specifically, (1) azoalkane decomposition proceeds with identical rates in solution and when covalently bound to a polymer matrix, (2) azoalkane homolysis produces a polymer chain cleavage, (3) the azoalkane: crosslinker ratio governs the rate of polymer cleavage, (4) the molecular weight:crosslink ratio equals the minimum MW obtainable, and (5) crosslinked polymer converts to a linear (soluble) material when the ratio of chain cleavages to crosslink equals 1:2. The above relationships should extend to other compounds and polymer types. Thus, it should be possible to tailor the thermal behavior of a degradable polymer by appropriate selection of azo structure. The exclusive literature on azo compounds homolysis kinetics[33] should facilitate the selection of compounds with favorable decomposition kinetics for a required temperature.

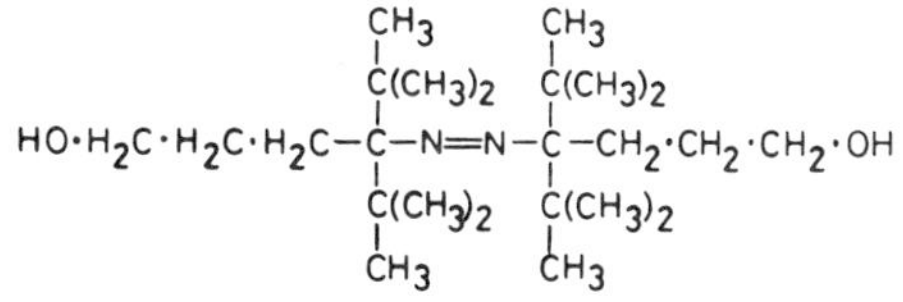

$$\mathrm{HO{\cdot}H_2C{\cdot}H_2C{\cdot}H_2C{-}\underset{\underset{\underset{CH_3}{|}}{C(CH_3)_2}}{\overset{\overset{\overset{CH_3}{|}}{C(CH_3)_2}}{C}}{-}N{=}N{-}\underset{\underset{\underset{CH_3}{|}}{C(CH_3)_2}}{\overset{\overset{\overset{CH_3}{|}}{C(CH_3)_2}}{C}}{-}CH_2{\cdot}CH_2{\cdot}CH_2{\cdot}OH}$$

BHPA

(78)

FIGURE 4.8

4.2.2. Type B: Chain End Scission

Block copolymers can be prepared using a polymeric initiator in a two-step procedure. Partial decomposition of the initiator groups in the presence of a first monomer results in a prepolymer containing initiator groups. This macroinitiator is decomposed in the presence of a second monomer forming block copolymers. The disadvantage of this method is that homopolymers are unavoidably formed and that the molecular as well as the chemical heterogeneity is large. However, the scope of this reaction is that all radically polymerizable monomers can be used; the blocks can as well be built up from any monomer mixture accessible to radical copolymerization. This statement is essentially limited only by solubility problems. The starting materials in both steps must be soluble to have a controlled reaction with high efficiency of the initiator and no crosslinking.

Polyesters containing azo groups can be easily obtained from 2,2′azobisisobutyronitrile and a diol using the Pinner synthesis. Polymerizing a monomer in the presence of the polyazoester for a duration of around one half-life of initiator decomposition leads to the formation of an azo group-containing prepolymer. These prepolymers are stable for years with respect to their initiator capability. During a reaction time of about ten half-lifes of initiator decomposition a second monomer (or monomer mixture) is polymerized with consumption of the remaining azo groups.

The structure of the resulting block copolymer is dependent upon the mode of termination of the monomers involved. If the polymerization of both the monomers is terminated by disproportionation, AB-block copolymers are formed. If one monomer gives rise to combination, the other dis-

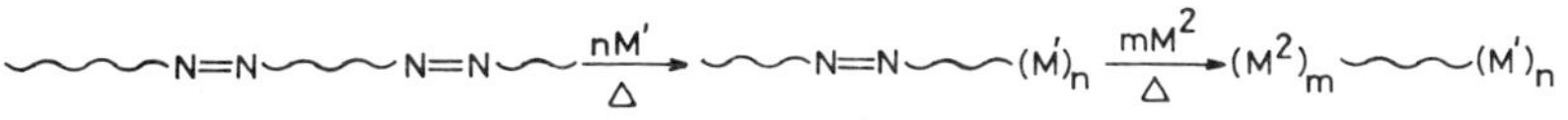

$$\sim\sim N{=}N\sim\sim N{=}N\sim \xrightarrow[\Delta]{nM'} \sim\sim N{=}N\sim\sim(M')_n \xrightarrow[\Delta]{mM^2} (M^2)_m\sim\sim(M')_n$$

FIGURE 4.9 Block copolymer formation by sequential initiators.

proportionation, mainly ABA or BAB block copolymers result. Multi-block copolymers are obtained if polymerization of both monomers is terminated by combination.

This possibility allows a wide variety of block copolymers to be formed as shown in the following examples. It is essential that the polymeric initiator is soluble in the medium for the first step. Although the polymeric initiator can also become soluble during the reaction, this process is hampered by a low efficiency and very often by crosslinking of the prepolymer. The crosslinking is probably caused by hydrogen abstraction from initiator-containing linear molecules and their reaction with other macroradicals.

Anand et al.[36] reported a series of polyazoesters which are soluble in most organic solvents. The rate of decomposition is comparable to AIBN. Azo groups decomposing inefficiently will react by combination. The efficiency was calculated by $P_n = v_M/(k/2 \cdot f \cdot v_I)$ with P_n: block length; v_M and v_I: amounts of monomer and initiator consumed in mol; k: number of independently growing chains forming one block; and f: efficiency. For example, for styrene the efficiency has been determined to be 0.3.

The preparation of prepolymers results in the formation of multiblocks if the corresponding monomer terminates by combination. Thermal cleavage of the azo groups in the prepolymer in the presence of a strong radical scavenger will give the isolated blocks. At high molecular weights, the quantitative reaction bears some concentration problems, but at $M < 10^{-5}$ the results clearly indicate azo group containing poly(1-vinyl-2-pyrollidone) are obtained by solution polymerization consisting of several blocks. The number of blocks is limited to the P_n of the polymeric initiator. The results indicate that the ratio of combination to disproportionation for vinyl pyrollidone is high. The block length can be regulated by the monomer/initiator ratio and temperature. The temperature must be adjusted in such a way that one half-life of initiator decomposition allows a reasonable conversion of the monomer. If the polymerization rate of a monomer is very high, good results can be obtained using a constant feed of monomer during the reaction.

Styrene is one of the monomers which has been polymerized with such initiators, and an azo polystyrene has been used as a prepolymer in block copolymer preparation with MMA. According to the two-stage scheme the polymerization has been performed in the temperature range of 70°C to 130°C.[37]

Other examples of macroinitiators[38] are polyazoesters accessible through the reaction of poly(alkylene oxide) such as poly(tetra ethylene glycol), and AIBN with acid catalysis.[38] The thermolabile azo groups allow the addition of new vinylic blocks into the polymer chain in association with the poly(alkylene oxide) ones. Partial decomposition of this polyinitiator in the presence of acrylamide results in a prepolymer containing azo groups. From GPC data, it has been concluded that termination of polymerization occurs predominantly by combination. The molecular weight increases with increasing monomer/initiator ratio. At the same time, higher reaction temperatures result in lower molecular weights. The

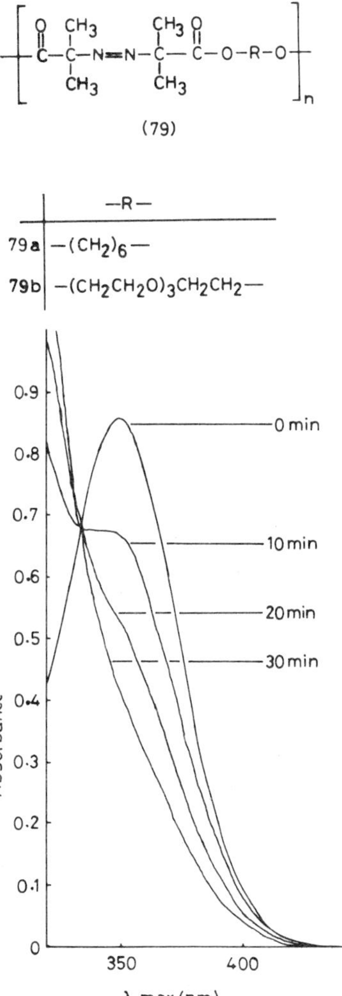

FIGURE 4.10 Decrease of the absorbance at 349 nm with time during the thermolysis of poly azo initiator in toluene at 85°C.

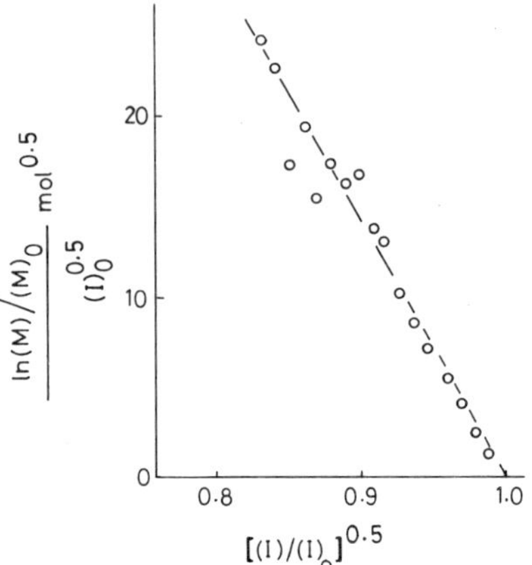

FIGURE 4.11 Conversion of 1-vinyl pyrollidone initiated with polyazoester. Cond.—60°C: $[M]_0 = 2.05$ mol/l; $[I]_0 = 4.13.10^{-3}$ mol/l: solvent-benzene.

study of the decomposition of the prepolymer shows a half-life at 70°C of 160 min compared to 310 nm for the polyazoester. The Arrhenius plot gives $k_d = 2.13.10^{10} \exp(-E/RT)$ with $E = 95.05$ kJ/mol.

It is also possible to combine hard and soft segments to synthesize blocks of different polarity and solubility. Based on similar principles, block copolymers formed from peptide and carbon-chain segments have been prepared.[39] The macroinitiators have been synthesized from bifunctional compounds containing labile azo groups such as azo bis 2,2′-diimidazoline isobutyric acid, and 2,2′ azo-bis-butyric acid hydrazide(80) as well as N-carboxyanhydrides of amino acids or styrene. With NCA, the results are a polypeptide macroinitiator having azo or S–S groups in the chain. With the second alternative, the reaction with styrene leads to a

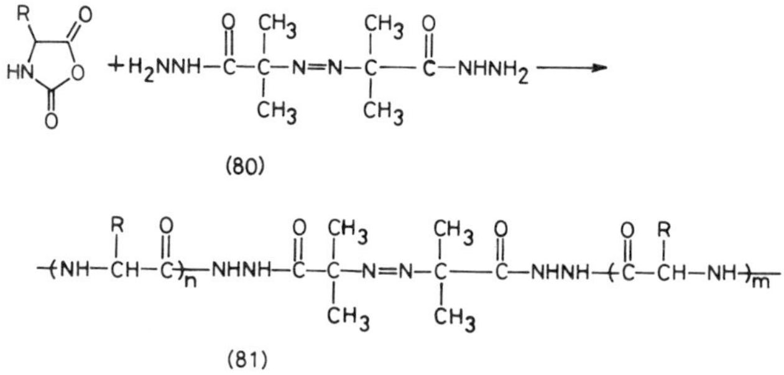

(80)

(81)

FIGURE 4.12

carbon-chain polymer initiator, with amino or hydrazide end groups. In order to obtain block copolymers, these macroinitiators are subsequently used in the polymerization of vinyl monomers or alternatively of the amino acid N-carboxyanhydride.[40]

These products lead to interest in ordered supramolecular and liquid crystal systems. Similar block copolymers with a polyvinyl block and polypeptide block were also prepared from other NCA compounds.[40,41]

One of the advantages of block copolymer synthesis by radical polymerization is to combine a wide variety of monomers. It is possible to put blocks of very different polarity or solution behavior together. The combination of hydrophilic/hydrophobic monomers will give interesting materials as membranes.

4.2.3. Type C: Side Chain Scission

Polymers containing side chain azo groups can be used as initiators for further polymerization and as prepolymers for production of graft copolymers.[42-45] The extent of grafting can be characterized in terms of grafting efficiency (ϵ):

$$\epsilon(\text{wt\%}) = \frac{(\text{weight of the grafted side chains})}{(\text{weight of the grafted side chains} + \text{weight of the homopolymer})} \times 100$$

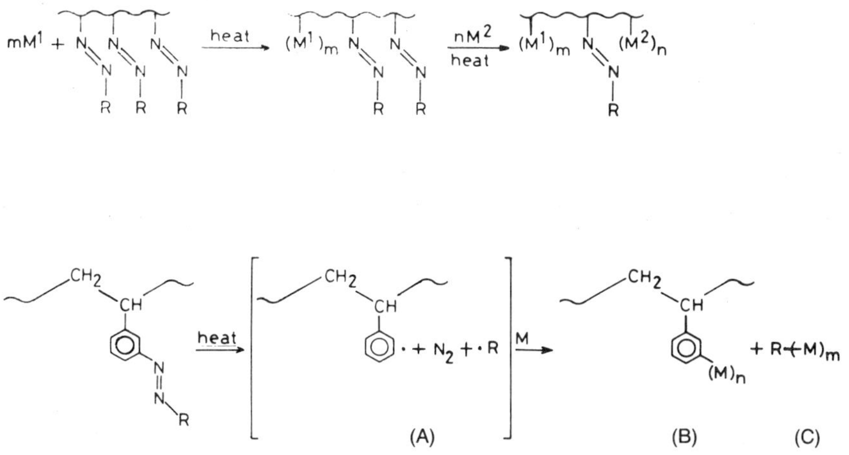

FIGURE 4.13 Mechanism of graft copolymer formation through thermolabile azo groups.

The experimental determination of ϵ involves separating the ungrafted backbone A, the graft copolymer B, and the homopolymer fraction C from the product mixture. The literature contains a variety of methods for achieving such a separation, e.g., selective extraction with the appropriate solvent/non-solvent mixtures,[46,47] and selective solubilization.[44] For each copolymer pair the appropriate solvent/non-solvent pair has to be found empirically and indeed the separations are often incomplete;[48,49] the solubility of the graft copolymer lies between those of the two homopolymers and depends on the extent of grafting which has taken place. A solution for these experimental problems would be to carry out the grafting reaction with a crosslinked polymer since the grafted product should be easily separated from any homopolymer formed simply by exhaustive extraction with a suitable solvent. The extent of grafting which has taken place should be easily obtained from the increase in the weight of original crosslinked polymer.

Gupta et al.[49] have reported the preparation of block and graft acrylamide-MMA polymer and block AN-MMA. The former was synthesized from chloromethylated PMMA by treatment with 2,2-azo diisobutyramidine or 2,2′ azobis 2-2(imidazoline-2-yl)propane) to give a polymeric initiator which on thermal activation reacted with acrylamide to give the graft copolymer. The same procedure was applied for the preparation of block AN-PMMA copolymer.

Macromolecular azo initiators have also been prepared from phenyl-(3-vinylphenylazo) sulphides by copolymerization reaction with butadiene and styrene.[41] They are accessible by coupling the corresponding diazonium salts with thiophenol and thionaphthol. Other macroinitiators of the polycarbonate class (82) having lateral azo groups have been studied by DSC method in order to determine the decomposition constants,[46] the activation energies, and the reaction enthalpy. The decomposition again has been found to follow first-order kinetics. The data demonstrates that the decomposition rate constants do not depend on the azo group content. The DSC technique has been applied too in the case of polystyrene having lateral side chain azo groups.[47] As opposed to polyazo carbonates, an increase in the azo group content of these initiators produces an increase of the decomposition rate constants, similar to the behavior of the low molecular weight compounds.

Polymer networks are formed by covalent, ionic and physical links between different polymer chains. One of the most important methods for the synthesis of networks is the chemical crosslinking of prepolymers containing carbon-carbon double bonds by adding a low molecular weight crosslinking agent. This agent has to be mixed with the polymer to be crosslinked. In order to avoid this additional mixing process, but also in order

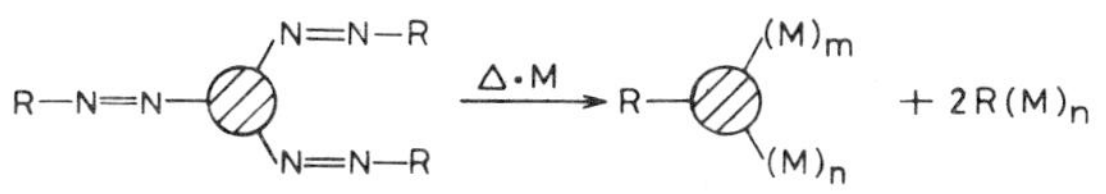

FIGURE 4.14

to guarantee a molecular distribution of the crosslinking agent in the polymer, Nuyken and coworkers[50] have developed a polymer system in which this agent is chemically linked to the polymer chain. For the synthesis of such polymers a mixture of butadiene/styrene (weight ratio 3:1) was used. Because of their structure, the azo monomers were presumed to react like styrene, so that the polymerization should also behave ideally. The thermal stability of these terpolymers was investigated by DSC measurements. The incorporation of the azo monomers into the terpolymers(83) was almost ideal under "cold rubber" conditions, meaning that the composition of the products does not change with conversion.

Linear azo polymers have been obtained via copolymerization of asymmetric azo monomers (84,85,86) as discussed in Chapter 3. The inclusion of the monomers (84) and (85) in a copolymer without decomposition of the azo group has been achieved via a redox-initiated emulsion polymerization technique.[50] Especially interesting in this respect are the systems that allow the azo monomer to be incorporated ideally into the polymer i.e., the azo monomer content of the polymer is the same as that of the monomer mixture and independent of the degree of conversion.

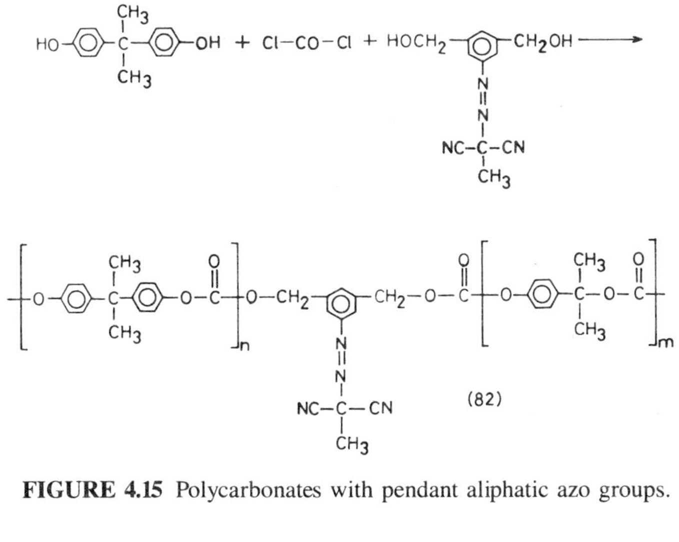

(82)

FIGURE 4.15 Polycarbonates with pendant aliphatic azo groups.

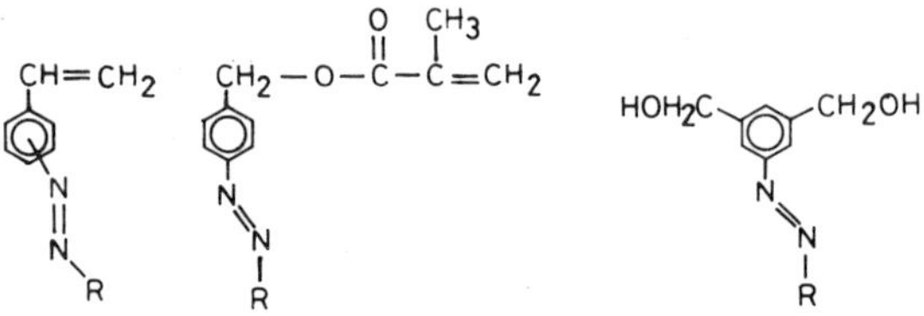

(84) (85) (86)

where R = $C(CN)_2CH_3$, $C(CN)(CH_3)_2$, $C(OOCCH_3)(R')_2$, SC_6H_5, and $R' = CH_3, C_6H_5$.

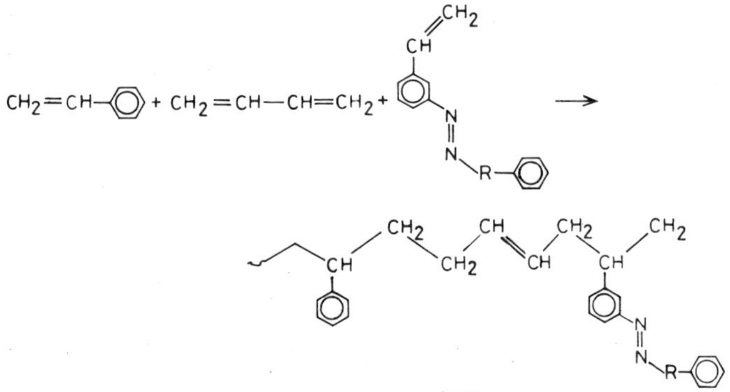

(83)

FIGURE 4.16 Aliphatic azomers for addition polymerization reaction.

With such systems it is possible to carry out the initial copolymerization to complete conversion, then raise the temperature of the emulsion, without breaking it up, to the decomposition temperature of the azo function, and then upon addition of new monomer, a graft copolymerization takes place. This technique avoids solubility problems and thus is particularly useful when large amounts of polymer are to be grafted with small amounts of monomer.

The polyazo compounds have also been applied in gelification processes of unsaturated polyester resins,[50] as hardening crosslinking agents for elastomers,[51] and natural and synthetic rubber blowing agents.[52-55]

4.3. Heat Resistant Polymers

It is generally recognized that the science of thermally stable materials entered a new phase in the mid-1950s.[53] For over 15 years, the emphasis in this new area was on the synthesis of new types of backbones. Numerous thermally stable aromatic and heterocyclic functional groups were incorporated into the polymers in an effort to find organic materials significantly more resistant to thermoxidation than those known. The ever present trade-off between properties and ease of product formation and processability continues to fuel further research in this area. The largest and the most important group of linear acyclic nitrogen polymers is polyamides.[54] The research in this area is due to the success of the aliphatic and later aromatic systems which have been developed into commercial products. Another contributing factor is undoubtedly the simplicity of the system and the plethora of routes to amides functions. The only monomers necessary are diacids and diamines or their derivatives. The amide groups in para aromatic polyamides are presumed to be in predominantly *trans* configuration.[54-58] The rotation barrier around the C–N bond is significant. The only bond rotation which could give rise to chain folding is thus inhibited. These polyamides are therefore made up of chain forming units, linked together in such a way that the chains are highly extended in the same direction, qualifying them to form optically anisotropic solutions in both amide and acid solvents. It has been shown that fibers spun from the anisotropic phase of these solutions have high modulus. As far as bond rotation is concerned, the structural considerations that apply to the amide group apply also to the azo group. Aromatic azo compounds exhibit *cis-trans* isomerism and the *trans* form is known to be considerably more stable than the *cis* form, which is directly related to the almost coplanar structure of the former compounds. Thus polyamides with azo-p-phenylene units have attracted a great deal of attention.

In addition, polymer chains which owe their rigidity to a structure consisting of planar rings connected by single links can have rod-like character while exhibiting greater solubility and lower melting temperature. Many such polymers are thermally stable as a result of their polyaromatic structure. The rod-like character of these polymers which can be altered by positional isomerism in the chain has a pronounced effect on the solution, and on the physical and thermomechanical properties of the polymers.

Fonton et al.[55] reported in their work, a possibility of introducing structural elements capable of conditioning the coloring features of the resultant polyamides. With this end in mind, they have prepared polyamides(87,88,89,90) from aromatic diacid chlorides and 3,3′diaminoazobenzene and 4,4′diaminoazobenzene. In general, polymers with a high degree of order and high resonance energy were obtained, and these characteristics were responsible for poor solubility but good thermal resistance. They have also reported other azo polymer types in order to study the influence which other different groups exert on the heat resis-

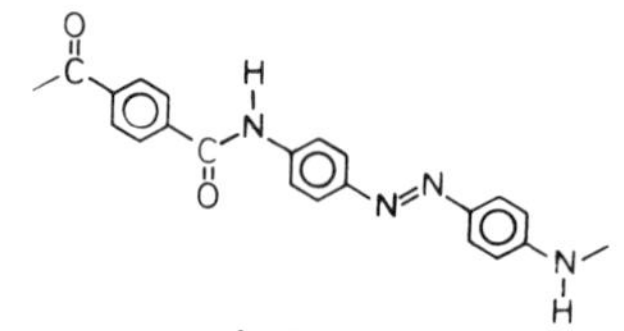

(87)

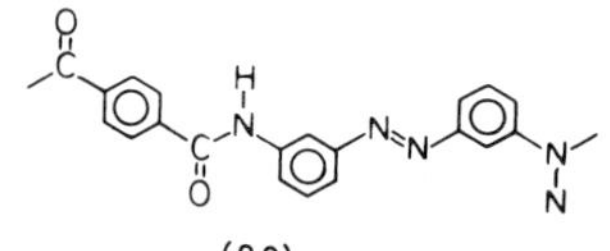

(88)

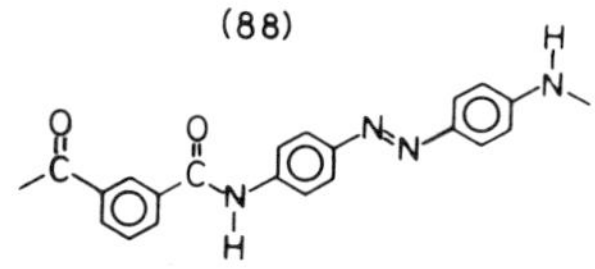

(89)

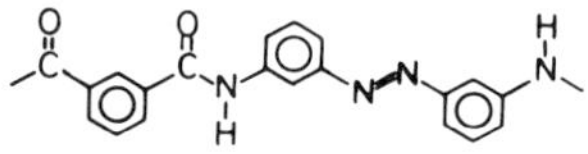

(90)

FIGURE 4.17 Thermally stable azoaromatic polyamides.

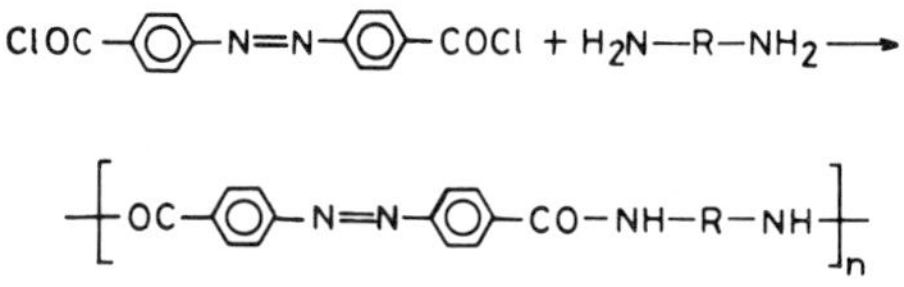

(91)

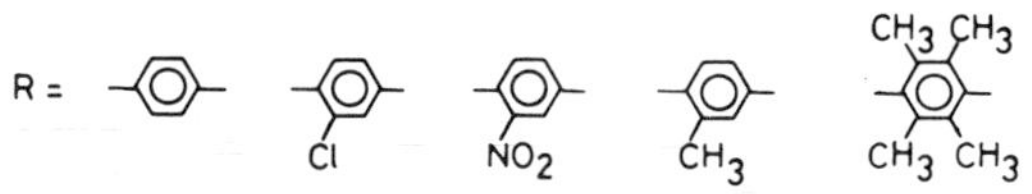

R =

PPDA PClPDA PNPDA PMPDA PTMPDA

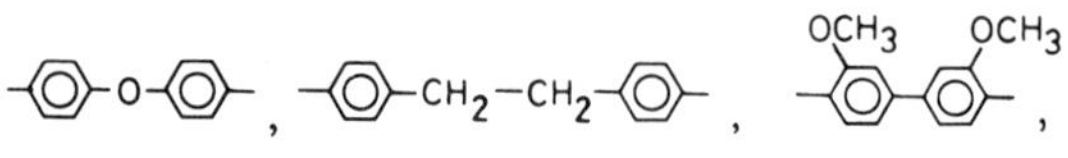

PEDA PDBA PODA

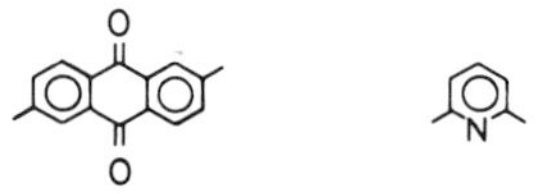

PAqDA PPDA

FIGURE 4.18 Azoaromatic polyamides.

tance and solubility to verify the action of the azo group on these polymeric structures. Polyurethanes, polyureas, polyesters and polycarbonates containing azo aromatic units were reported. One feature which distinguishes these polymers is their high resistance to hydrolysis.

Nanjan et al.[59-64] have reported the synthesis of several azopolyamides(**91**) and azopolyhydrazides. TGA curves of various polyamides investigated are shown in Figure 4.19. The yields and viscosities of polymers prepared by the low temperature solution method are higher than those prepared by the interfacial polycondensation method. These polymers showed gradual weight loss in two steps. Rapid weight loss occurred for most of the polymers in the 315–545°C range corresponding to a 20% weight loss. Hence, the thermal stabilities are compared at the temperature at which a 20% weight loss takes place.

A TGA curve of PODA showed the weight loss in air and nitrogen were comparable, indicating that the mechanism of decomposition is nearly the same in both the conditions. All the polymers exhibited an exotherm indicating decomposition without melting. However, PEDA containing flexible ether linkages showed endotherm, which might relate to melting.

Polyhydrazides(**92**) are of interest as precursors to various heat resistant and chemically stable polyoxadiazoles.[65] It has also been shown that paralinked polyhydrazides form liquid crystalline solutions in aqueous organic bases and sulfuric acid. All five polyhydrazides investigated showed a change in intensity of color when strong bases like 2M NaOH and 25% aq. ammonia were added.[62] The original color was restored when dilute

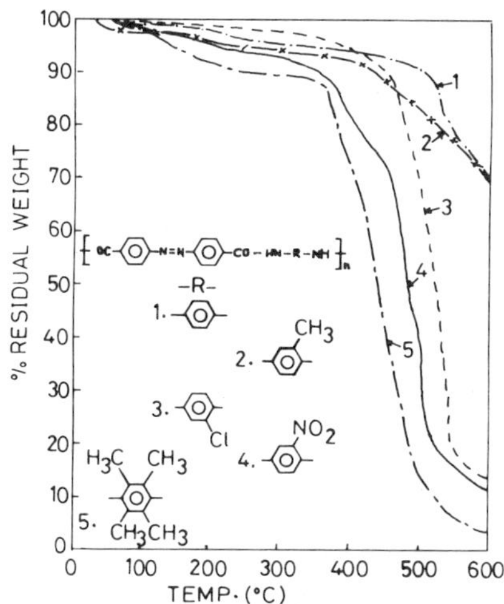

FIGURE 4.19 TGA curves of azopolyamides in N_2.

acids were added. This is due to the keto-enol tautomerism (Figure 4.20) exhibited by the hydrazocarbonyl units. Enolization can conceivably result in resonance interactions that involve the azo group. A typical example of enolization is shown in Figure 4.20.

The polyhydrazides enolized with aqueous ammonia as well as NaOH showed bands characteristic of a C=N group around 1640 cm^{-1} and 1490 cm^{-1}. DTA studies showed that the resonance stabilized polyhydrazides had greater resistance to cyclodehydration then the other polyhydrazides groups. This was explained on the basis of the less closely packed structure of these polymers due to the presence of the meta-linked aromatic rings. The TG curves showed 5–10% of the weight loss for all polymers within 200°C due to expulsion of the absorbed moisture. The first significant break corresponding to the beginning of the cyclodehydration reaction occured in the range of 220–390°C. It was found that the rate of cyclodehydration proceeded at a faster rate for enolized polyhydrazides than for other polyhydrazides. These cyclodehydration results were explained as due to the formation of adjacent oxadiazole rings. The polyoxadiazoles started to decompose gradually after 300°C.

New silicone-containing azopolyamides(93) (Figure 4.21) have been prepared from 4,4″ diamino azobenzene and four structurally related organosilicon acid chlorides.[66] The polymers were yellow to brown in color. On examining the inherent viscosities of aramids, it was found that the molecular weights are much more dependent on the reactivity of the acid dichloride. The aramid obtained from 4-DMCS has the highest viscosity while that from 3-DPCS has the lowest viscosity. In order to study the effect of structures on the thermal stability of the polyamides, the thermoxidative degradation was studied by TG, DTA and DTG curves for all the polymers, which showed more or less similar patterns.

A comparison of thermal characteristics shows that polymers obtained from 4-DPCS has higher thermal stability due to symmetry and its more aromatic structure. The polymer derived from 3-DPCS is thermally least stable because of its low molecular weight. Polymers from 3-DMCS and 4-DMCS showed appreciable thermal resistance.

Through the catalyzed oxidative coupling of primary aromatic diamines a great variety of aromatic azo polymers have become easily accessible[64] and investigations of their properties in bulk as well as fabricated form attracted attention.

All aromatic azopolymers are colored owing to the strongly chromophoric azo group. Even the azo block copolymer derived from a phenyl oxide-isophthalamide backbone which has only one group per molecular weight of 3500 is bright yellow. Naturally, the shade of color of individual polymers depends on the structure of the repeating unit. Fully conjugated polymers such as (Ar–N=N–) are essentially black.

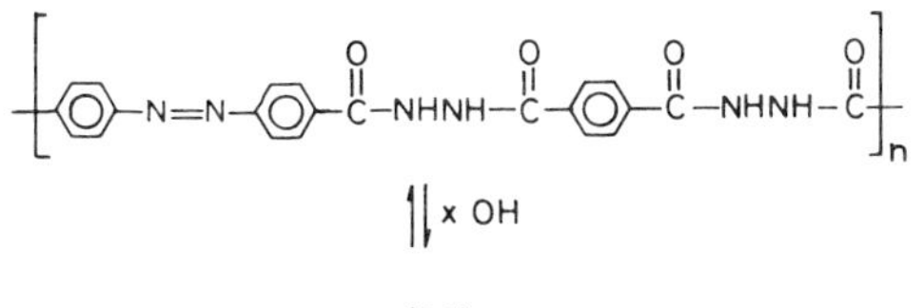

x OH

(92)

FIGURE 4.20 Keto-enol equilibrium in poly azo hydrazides.

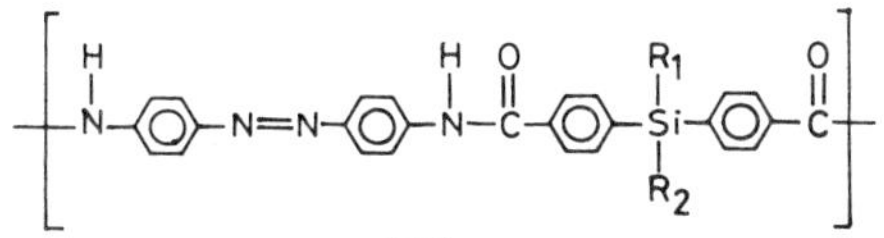

(93)

where $R_1 = R_2 = -CH_3$ or $-C_6H_5$

FIGURE 4.21

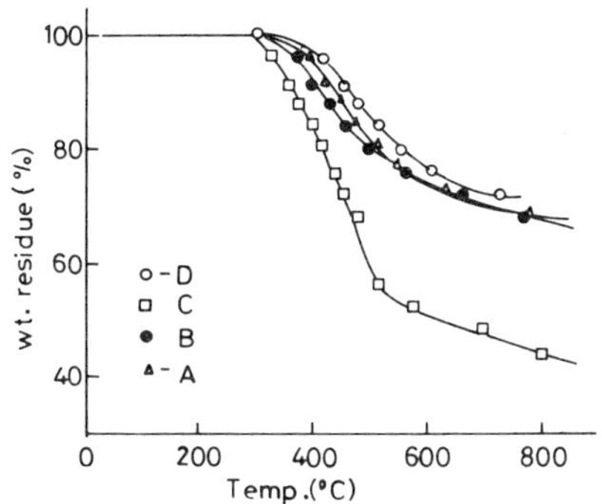

FIGURE 4.22 TGA curves in air at 10°C/min for polyamides(93).

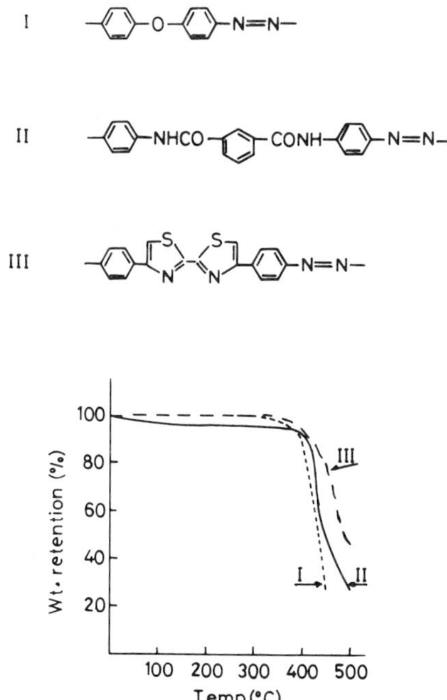

FIGURE 4.23 TGA curves of samples I, II, III in air. Heating rate 15°C/min.

Aromatic azo polymers also show a high degree of crystallinity without annealing[65] as evidenced by X-ray diffraction patterns. The high degree of crystallinity of poly(azo phenylene oxide) (PPO) is surprising in view of the fact that other polymers such as PPO and polysulfone which have a predominance of aromatic ether linkages in the backbone show little or no crystallinity unless they are annealed for several days in a suitable solvent.

Figure 4.23 shows TGA diagrams of polymers in air. The higher stability of I in Figure 4.23 in air than in nitrogen atmosphere was explained by a weight gain arising from oxidation, thus compensating for some of the weight loss caused by expulsion of N_2.[66] These polymers were stable to

Table 4.3 Standard Tensile Data of an Azo Polymer Fiber.

	As Spun	Hot Drawn
Tenacity, grams/denier	2.5	4.4
Elongation, %	19.6	9.0
Tensile modulus, grams/denier	58.0	93.0

Table 4.4 Tensile Properties at Varying Temperatures.

	Fiber Temperature (°C)				
	250	300	350	400	450
Tenacity, grams/denier	1.8	1.4	1.5	1.4	0.6
Elongation, %	5.7	8.7	6.6	4.8	1.4
Tensile modulus, grams/denier	47	37	39	40	47

more than 300°C, but decomposed with a catastrophic weight loss (> 10%) in the range of 360–400°C. Rapid thermal degradation was always accompanied by a strong exotherm. This exotherm combined with the narrow range (360–400°C) of its occurrence irrespective of polymer structure indicates that the first step in the degradation is the elimination of the azo group as molecular nitrogen.[67] Fully aromatic azopolymers described showed no melting or softening up to the temperature of thermal degradation.

Bach et al.[69-71] investigated the poly(isophthalamide) of 4,4′-diamino azobenzene (Figure 4.23, II) for its fiber-forming properties of aromatic azopolymers. Table 4.3 shows standard tensile data of the "as spun" as well as the hot drawn (1.5 × 350 C) fiber. As shown, this polymer has tensile properties fairly typical of aromatic polyamides.

In view of the good stability of the aromatic azopolymers at elevated temperatures, a fair degree of retention of tensile properties was observed up to about 400°C (Table 4.4).

The above polymers also show the retention of tenacity compared with another aromatic polyamide, M3P, the polytherephthalamide of bis (m-aminobenzoyl) m-phenylene diamine. Fibers of polymer II show a comparatively stronger resistance to light degradation. These values compare favorably with the strength of a commercial aromatic polyamide.

4.4. Conclusion

Numerous patents and publications describe possible industrial uses for azo polymers as thermolabile and thermostable polymers. Although several publications describe aliphatic azo-containing polymers as sources of macroradicals in the preparation of block and graft copolymers, no industrial processes have been reported thus far. In comparison with polymeric peroxides as polymeric initiators, interest in using polymeric azo compounds as initiators has been low. The literature reports only those structures which incorporate AIBN moiety as the repeating unit and very little of other related structural variations have been investigated.

Incorporation of azoaromatic units into rigid-chain polymers to enhance thermal stability is again limited to azobenzene units. No other azoarenes such as azopyridines, azonaphthalenes etc. have been incorporated. As a result, I am constrained to make any statements on the general thermal behavior of these polymers. But these gaps pointed out may also stimulate a greater diversification of thought by those already involved in the field, and perhaps bring new ideas and solutions to related areas where considerable problems still remain to be solved.

4.5. References

1. Nuyken, O. and R. Weidner. 1986. "Graft and Block Copolymers via Polymeric Azo Initiators," *Adv. Polym. Sci.* 73/74:145.
2. Nuyken, O. 1985. In *Encyclopedia of Polymer Science & Engineering, 2nd Ed.*, New York, NY: Wiley & Sons, 2:158.
3. Critchley, J. P., G. J. Knight and W. W. Wright. 1983. *Heat-Resistant Polymers*, New York, NY: Plenum Press, p. 200.
4. Cassidy, P. E. 1980. *Thermally Stable Polymers*, New York, NY: Marcel Dekker Inc.
5. Hergenrother, P. M. 1985. In *Encyclopedia of Polymer Science & Engineering*, New York, NY: Wiley & Sons, 7:639.
6. Lewis, F. M. and M. S. Matheson. 1949. "Decomposition of Aliphatic Azocompounds," *J. Amer. Chem. Soc.*, 71:747.
7. Flory, P. J. 1957. *Principles of Polymer Chemistry*, New York, NY: Cornell University Press, p. 119.
8. Shah, N. A., F. Leonard and A. Tobolsky. 1951. "Phthaloyl Peroxide as a Polymerization Initiator," *J. Polym. Sci.*, 7:537.
9. Tsvetkov, N. S. 1961. "Polymerization Kinetics of Styrene in the Presence of Phthaloyl Peroxide," *Vysokomolek. Soedin*, A3:408.
10. Tsvetkov, N. S. and R. F. Markovskaya. 1964. "Use of Polymeric Peroxide of Sebacic Acid in the Synthesis of Polystyrene and Block Copolymers," *Vysokomolek. Soedin.*, A6:2051.
11. Tsvetkov, N. S. and R. F. Markovskaya. 1965. "High Conversion Polymerization of Styrene and MMA in the Presence of Polymeric Peroxides," *Vysokomolek. Soedin.*, A7:169.
12. Tsvetkov, N. S. and Ye Beletskaya. 1967. "Polymeric Peroxides as Initiators for Radical Polymerization," *Ukr. Khim. Zh.*, 33:380.
13. Tsvetkov, N. S. and R. F. Markovskaya. 1974. "Polyfunctional Initiators for Radical Polymerization and Molecular Weight Distributions of Polymers," *Vysokomol. Soedin.*, A16:1936.

14. Smets, G. and A. Woodward. 1954. "Preparation of Block Copolymers," *J. Polym. Sci.*, 14:126.

15. Simionescu, Cr. I., E. Comanita, M. Pastravanu and S. Dumitriu. 1986. "Progress in the Field of Bi- and Poly-Functional Free-Radical Polymerization Initiators," *Prog. Polym. Sci.*, A. D. Jenkins and V. T. Stannett, eds., 109:1.

16. Walz, R. and W. Heitz. 1978. "Preparation of $(AB)_n$–Type Block Copolymers by the Use of Azo Polyesters," *J. Polym. Sci. Polym. Chem. Ed.*, 16:1807.

17. Walz, R., B. Bomer and W. Heitz. 1977. "Monomeric and Polymeric Azo Initiators," *Makromol. Chem.*, 178:2527.

18. Heitz, W., M. Lattekamp, C. Oppenheimer and P. S. Anand. 1983. "Synthesis of Block Sequences by Radical Copolymerization," *ACS Symposium Series, Washington DC, ACS*, 212:337.

19. Dicke, H. R. and W. Heitz. 1981. "Synthesis and Characterization of Surface-Active Azo-Group Containing Polyacrylamides," *Makromol. Chem. Rapid Commu.*, 2:83.

20. Dicke, H. R. and W. Heitz. 1982. "Surface-Active Azo-Group Containing Polymers for Emulsion Polymerization," *Colloid Polym. Sci.*, 260:3.

21. George, M. H. and J. R. Ward. 1973. "Polymerization of Styrene by Poly-(bisphenol A 4,4′-azobis (4-cyanopentanoate)," *J. Polym. Sci. Polym. Chem. Ed.*, 11:2909.

22. George, M. H. 1967. Chapter 3 in *Vinyl Polymerization*, G. E. Ham, ed., New York, NY: Marcel Dekker, Vol. 1.

23. George, M. H. 1964. "Polymerization of Styrene in N,N′-Dimethyl Formamide," *J. Polym. Sci.*, A2:3169.

24. George, M. H. and J. R. Ward. 1973. "Polymerization of Styrene in N,N′Dimethyl Formamide-II," *J. Polym. Sci. Polym. Chem. Ed.*, 11:377.

25. Sheppard, C. S. and R. E. MacLeay. B.R.D. 1,940.473, Feb. 10, 1973. Chem. Abstr. 72,90983m, 1970.

26. MacLeay, R. E. and C. S. Sheppard. U.S. Pat. 3,987,024, 1976.

27. MacLeay, R. E. and C. S. Sheppard. U.S. Pat. 4,075.286, 1978. Chem. Abstr. 88, 170790s, 1978.

28. Nelson, S. F. and P. D. Bartlett. 1966. "Azocumene I:Preparation and Decomposition of Azocumene," *J. Amer. Chem. Soc.*, 88:137.

29. Smith, D. A. 1967. "The Thermal Decomposition of Azonitrile Polymers," *Makromol. Chem.*, 103:301.

30. Overberger, C. G., J. Anselme and J. R. Hall. 1963. "Azo Compounds. Dipole Moments and Spectral Data," *J. Amer. Chem. Soc.*, 85:2752 and preceding papers.

31. Kenley, R. A. and G. E. Manser. 1985. "Degradable Polymers Incorporating a Difunctional Azo Compound into a Polymer Network to Produce Thermally Degradable Polyurethanes," *Macromolecules*, 18:127.

32. Koenig, T. 1973. In *Free Radicals*, J. A. Kochi, ed., New York, NY: Wiley, 1:119.

33. Hill, J. W. U.S. Pat. 2,556,876, 1951.

34. Furukawa, J., S. Takamori and S. Yamashita. 1967. "Preparation of Block Copolymers with a Macroazonitrile as an Initiator," *Agnew. Makromol. Chem.*, 1:92.

36. Anand, P. S., H. G. Stahl, W. Heitz, G. Weber and L. Bottenbruch. 1982. "Block Copolymers by Radical Polymerization," *Makromol. Chem.*, 183:1685.
37. Krakovyak, M. G., E. V. Anufrieva and S. S. Skorokhodov. 1972. "Synthesis and Copolymerization of Anthracene-Containing Polymers," *Vysokomolek. Soedin.*, A14:1127.
38. Jimura, K. and N. O. Koide. 1982. *Rhe. Proc. IUPAC, Makromolek. Symp.*, 28th. 825; Chem. Abstr. 98, 159128p, 1983.
39. Vlasov, G. P. and G. D. Rudskovskaia. 1980. "Synthesis of Block Copolymers Containing Carbon Chain and Poly(aminoacid) Blocks," *Vysokomolek. Soedin.*, B22:216.
40. Billot, J., A. Douy and B. R. M. Gallot. 1976. "Synthesis and Structural Study of Block Copolymers with a Hydrophobic Polyvinyl Block and a Hydrophilic Polypeptide Block," *Makromol. Chem.*, 177:1889.
41. Billot, J., A. Douy and B. R. M. Gallot. 1977. "Preparation, Fractionation and Structure Determination of Block Copolymers of Polystyrene and Polypeptide Block," *Makromol. Chem.*, 178:1641.
42. Nuyken, O., M. Dorn and R. Kerber. 1979. "Azo Initiators–X," *Makromol. Chem.*, 180:1651.
43. Kerber, R., O. Nuyken and R. Steinhausen. 1976. "Azo Initiators–III: Effect of Solvents on the Thermal Degradation of p-Methoxy Phenyl Azo-2-Methyl Propanedinitrile," *Makromol. Chem.*, 177:1357.
44. Kerber, R., O. Nuyken and R. Steinhausen. 1977. "Grafting and Crosslinking with Copolymers Containing Azo Groups," *Makromol. Chem.*, 178:1833.
45. Kerber, R., O. Nuyken and M. Dorn. 1978. "Azo Initiators–VII," *Makromol. Chem.*, 179:1803.
46. Fuchs, O. 1950. "Fractionization of High Polymers," *Makromol. Chem.*, 5:245.
47. Fuchs, O. 1951. "Fractionization of High Polymers by Solution Method," *Makromol. Chem.*, 7:259.
48. Fischer, J. P. 1975. "Kinetics and Morphological Studies on the Free Radical Grafting of Styrene on Polybutadienes," *Angew. Makromol. Chem.* 33:35.
49. Gupta, S. N. and U. S. Nandi. 1975. "Preparation of Graft and Block Copolymers by Means of Amide Linkages," *Makromol. Chem.*, 176:3179.
50. Nuyken, O., J. Gerum and R. Kerber. 1980. "Grafting of Azo Group Containing Polycarbonates," *Angew. Makromol. Chem.*, 91:143.
51. Heitz, W., H. G. Stahl and R. Dicke. Ger. Off. 3,005,889, Mar. 9, 1981, Chem. Abst., 151474v, 1981.
52. Sheppard, C. S. and R. E. Macleay. U.S. Pat. 3,755,443, Aug. 28, 1973. Chem. Abst. 80,4063x, 1974.
53. Bach, H. C. and W. B. Black. 1969. "Aromatic Azopolymers Produced by Oxidative Coupling of Primary Aromatic Diamines," *Adv. Chem. Ser.*, 91:679.
54. Stille, J. K. 1974. In *Proc. Intl. Symp. Macromol*, E. B. Mano, ed., New York, NY: Elsevier, p. 95.
55. Fonton, J., S. G. Babe and J. De Abajo. 1974. In *Proc. Intl. Symp. Macromol.*, E. B. Mano, ed., New York, NY: Elsevier, p. 305.
56. Morgan, P. W. 1977. "Synthesis and Properties of Aromatic and Extended Chain Polyamides," *Macromolecules*, 10:1381.

57. Kwolek, S. L., P. W. Morgan, J. R. Schafgen and L. W. Gulrich. 1977. "Synthesis of Anisotropic Solutions and Fibers of Poly(1,4-benzamide)," *Macromolecules,* 10:1390.
58. Preston, J. 1978. Chapter 4 in *Liquid Crystalline Order in Polymers*, A. Blumstein, ed., New York, NY: Academic Press, p. 141.
59. Balasubramanian, M., M. J. Nanjan and M. Santappa. 1979. "Synthesis of Polyamides Containing Azo Group in the Polymer Chain," *Makromol. Chem.*, 180:2517.
60. Nanjan, M. J., M. Balasubramanian, K. S. V. Srinivasan and M. Santappa. 1977. "Synthesis of Poly(2,2-bis[4{p-aminophenoxy}phenyl] propane-terephthalic acid)," *Polymer*, 18:411.
61. Balasubramanian, M., M. J. Nanjan and M. Santappa. 1981. "Synthesis of Polyamides Containing 4,4′-azodibenzamido Units," *Makromol. Chem.*, 182:853.
62. Jayaprakash, D. and M. J. Nanjan. 1982. "Synthesis and Characterization of Certain Polyhydrazides Containing Azo Group in the Main Chain," *J. Polym. Sci. Polym. Chem. Ed.*, 20:1961.
63. Balasubramanian, M., M. J. Nanjan and M. J. Santappa. 1982. "Synthesis, Characterization and Fiber Studies of Certain Aromatic Polyamides," *Appl. Poly. Sci.*, 27:1429.
64. Jayaprakash, D., M. Balasubramanian and M. J. Nanjan. 1985. "Synthesis and Characterization of Certain Polyamides Containing the Azo Group in the Main Chain," *J. Polym.Sci. Polym. Chem. Ed.*, 23:2319.
65. Frazer, A. and F. T. Wallenberger. 1964. "Aliphatic Polyhydrazides-Low Temperature Solution Polymerization," *J. Polym. Sci.*, A2:147.
66. Jadhav, J. Y., N. N. Chavan and N. D. Ghatge. 1985. "Preparation of Silicone-Containing Polymers – III," *Eur. Polym. J.*, 20:1009.
67. Bach, H. C. and W. B. Black. 1969. "Aromatic Azopolymers Produced by Oxidative Coupling of Primary Aromatic Diamines," *J. Polym. Sci.*, C22:799.
68. Bach, H. C. and H. E. Hinderer. 1970. "Highly Ordered Azoaromatic Polyamides through a Novel Dimerization by Oxidative Coupling," *Polym. Prep. Amer. Chem. Soc. Div. Polym. Chem.*, 11:334.
69. Bach, H. C. U.S. Pat. 3,501,444, 1970.
70. Bach, H. C. U.S. Pat. 3,598,768, 1971.
71. Bach, H. C. U.S. Pat. 3,732,200, 1973.
72. Johnson, R. N. 1967. "Thermal Stability of Poly(aryl ethers) Prepared by Aromatic Nucleophilic Substitution," *J. Polym. Sci.*, A-1, 5:2375.
73. Karasz, F. E. and J. M. O'Reilly. 1965. "Thermal Properties of Poly(2,6-dimethyl Phenylene Ether)," *Polym. Letters*, 3:561.

5

PHOTOCHEMISTRY OF AZO FUNCTIONAL POLYMERS

5.1. Introduction

Azo compounds are also photosensitive and they absorb ultraviolet radiation in relatively high quantum yield. Depending on the substitution on the $-N=N-$ moiety, these compounds undergo either photolysis or photoisomerization or both. Aliphatic azo compounds undergo photolysis accompanied by isomerization and produce free radicals. However, the quantum yield for radical generation is low because the absorbed radiant energy primarily excites and isomerizes ground state *trans* azo compounds to excited state *cis* and *trans* azo structures, which are partly deactivated to the ground state and partly decomposed to free radicals and nitrogen. Most radicals are generated from the *cis* forms, which are thermally less stable than the *trans* forms. Although many studies have been carried out on the photodecomposition of azo compounds,[1,2] commercial aliphatic azo compounds are used primarily as thermal sources of free radicals. The extension of the photolysis of the aliphatic azo group for the design of photolabile polymers has not been reported.

The characteristic reaction of aromatic azo compounds is the photoinduced *trans-cis* isomerism. In fact, all the reports on the photochemical behavior of azo-containing polymers relate to incorporation of azobenzene units in polymeric structure and their photo-induced isomerization. The aromatic azo compounds are photostable and undergo isomerization without any side reactions. Only a small part of the numerous photochemical reactions in organic compounds result in reversible structural change and azobenzene is among the most studied along with stilbenes and spirobenzopyrans. These isomerizations are accompanied by a change in physical properties, in particular, electronic spectral properties. It has tempted several groups of workers to try to use this photochromic effect in various polymer systems for controlling and regulating light fluxes and for data recording etc.

The challenge in polymer chemistry is the application of fundamental physical and chemical problems to long chain macromolecules which may coil, branch, be chemically crosslinked or take part in other orders of structural complexity. For example, the photochemical isomerizations which occur in small molecules can also be induced to occur in macromolecules. But, in a macromolecular environment there are constraints which are not present in molecular photochemistry, and hence we can learn a great deal by the study of such polymer systems. The progress made in macromolecular photochemistry indicates that the dominant factor is the free volume of the system. The free volume available to a reactive site or dissolved probe controls the course of photophysical and photochemical processes. Likewise molecular mobility plays an important role in determining the course of photochemical reactions in polymers and is related to the size of the molecule, the flexibility of the polymer chain and whether the polymer is in solution or solid state. The details of functional group transformations attached to polymer chains and the factors influencing these changes are nowhere more evident than in the photochemistry of azo functional polymers.

5.2. Photochemistry of Azobenzene: *Cis* ⇌ *Trans* Isomerism

Azobenzene is a textbook representative for the phenomenon of *cis-trans* isomerism demonstrating the rotoresistant property of the N=N double bond. In fact, a number of theoretical and experimental papers indicate interest in the isomerization mechanism of azobenzene.[3-5] Spectroscopists and photochemists have been aware of the inversion-rotation problem for a long time and many experimental papers deal with this subject.[6-10]

Photoinduced isomerism of azobenzene also proceeds with large structural change as reflected in the dipole moment and change in geometry. The isomerization involves a decrease in the distance between the para carbon atoms in azobenzene from about 9.0 Å in *trans*-form to 5.5 Å in *cis*-form and the local contraction may be even longer.[11] Likewise, normal azobenzene has no dipole moment while the non-planar *cis* compound has 3.0 D.[11] These properties, if manifested in polymers, are of utmost value in probing conformational dynamics of macromolecules by site-specific photolabelling, in estimating the free volume in crosslinked networks and in designing photoreactive polymers responsive to external stimuli.

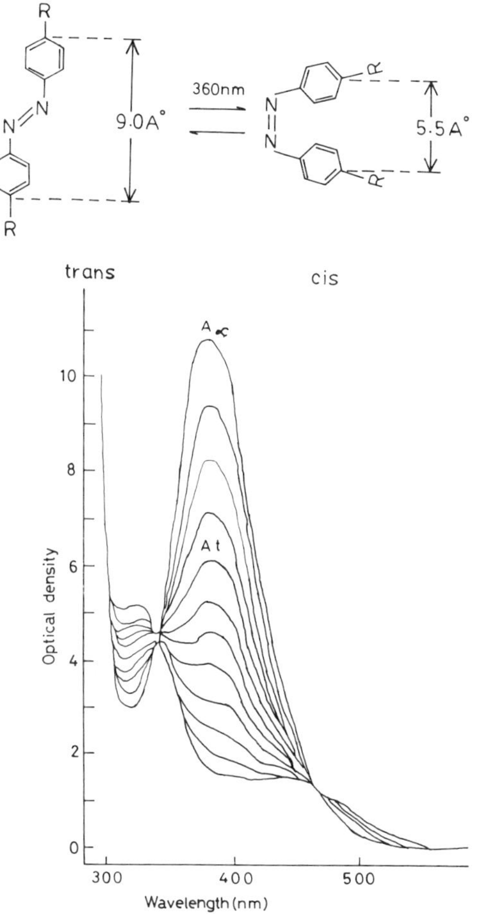

FIGURE 5.1 Absorption spectra of azobenzene in $CHCl_3$ showing thermal recovery (Temp 28°C) and accompanying geometrical changes.

5.3. Macromolecular Photochemistry

Much of the work done in macromolecular photochemistry involving azobenzene moieties has progressed along two lines. In earlier studies[12,13] touched upon in this survey, the motivation was a desire to clarify a physical problem. What is the nature of hindered rotation in long chain molecules? What is the distribution of free volume in the glassy state? What is the flexibility of chain molecules? Furthermore, azo labels can be selectively attached to the side chains, main chain, crosslinks or chain ends of the polymer as photochromic "probes." Will the selective site-labelling allow one to identify motions associated with the specific location of the polymer chain? By varying the molecular structure and the polymeric system, as well as by inhibiting the motion of the polymer matrix, is it possible to construct materials having photochromic characteristics? These factors together with improved technology for obtaining photochromic materials rather than introducing a low molecular weight compound in a polymeric matrix, explains the increasing interest among the research workers.

(Already in 1911 Ciamician had described the color changes some substances undergo under the influence of light. Included in his historical address were the following words ". . . Phototropic [now photochromic] substances, which often assume intense colors in the light, and afterwards return in the darkness to the primitive color might be used very effectively. . . . The dress of a lady so prepared would change its color according to the intensity of the light. . . . conforming automatically to the environment, the last word of fashion for the future"–Ciamician, G. 1912. *Science*, 36:385.)

As the needs for polymer materials are changing from structural materials to functional materials, the current trend is to employ azobenzene moiety as a "trigger" to induce morphological changes which can be used for light-driven chemical valves, light induced porous filters, etc. Chemical substances which exhibit photoinduced structural changes are the candidates not only of chemical condensers for the storage of light energy but also of mediators of light energy to chemical functions. The possibility thus arises that conformations of azobenzene-containing polymers can be controlled by photoinduced configurational changes of azobenzene so that they may provide desired chemical functions. Expectedly, this would lead to the possibility of controlling the chemical functions by an "on-off switch."

5.3.1. Dynamics through Kinetics: Azobenzene as a "Probe"

The investigation of the chemical and physical behavior of the photochromic moieties in polymeric matrices and the influence of polymers on photochromism is a relatively recent phenomenon. While Flory's[14] demonstration that the attachment of a functional group will, in many cases, not lead to an altered activity, a number of interesting effects may be observed in special cases.[15] Differences in reaction rates were expected and found at the ends of polymer chains, along the chain and at irregularities within the matrices. From the literature, it is obvious that the nature and morphology of polymer has an essential influence on the photo-

and thermochromism of a chromophore in a given polymer matrix.[12,13] The size of a chromophore, its shape and its configuration and its point of attachment to a polymer chain also play a role in determining the effective free volume available to an isomerizable group. A comparison of results of different authors is difficult because the systems investigated and the experimental conditions employed are too different; besides, no meaningful theories exist as to how photochromic processes are linked to polymer properties. The local arrangement of the polymeric environment, i.e., macroscopic and microscopic values of polarity, viscosity, T_g and tacticity as well as the history of the sample can affect photochemical or photophysical events.

A. Backbones – "Crankshaft-Like Motion"

It is customary to consider a *cis-trans* isomerization around a double bond as a chemical process, while hindered rotation around a single bond is thought of as a physical transition between two states of the same molecule. Still, the distinction between these two kinds of processes is of no fundamental significance, having been made because the most typical isomerizations around a double bond are very slow compared to typical rotations around single bonds, so that *cis-trans* isomers are usually easy to separate, while isomers involving rotation around single bonds have been separated only in exceptional cases. While the conformational distribution in flexible chain molecules has been a subject of intensive theoretical and experimental study ever since Kuhn's pioneering work,[16] the dynamics of conformational transition have become only recently a subject of experimental study. The belief is widely held that conformational transitions in the backbones of long polymer chains must involve two correlated hindered rotations (a crankshaft-like motion) so as to avoid the need for a long chain segment to swing through the viscous solvent medium (Figure 5.2).[17] If this model represents correctly the physical reality, the free en-

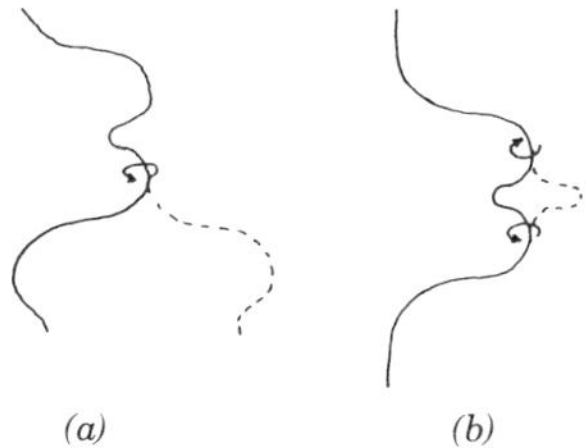

FIGURE 5.2 Schematic representation of conformational transition in a flexible chain molecule: (a) rotation around a single bond, (b) correlated rotation around two bonds.

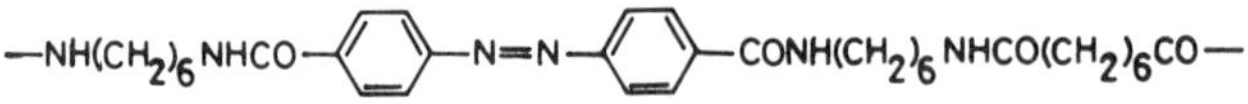

$-NH(CH_2)_6NHCO-C_6H_4-N{=}N-C_6H_4-CONH(CH_2)_6NHCO(CH_2)_6CO-$

(94)

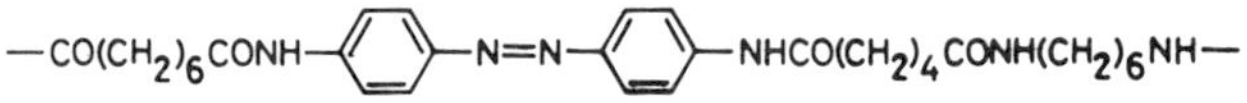

$-CO(CH_2)_6CONH-C_6H_4-N{=}N-C_6H_4-NHCO(CH_2)_4CONH(CH_2)_6NH-$

(95)

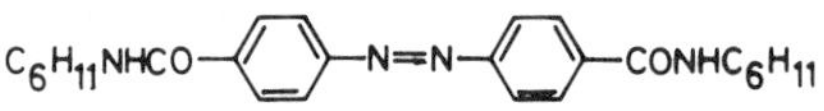

$C_6H_{11}NHCO-C_6H_4-N{=}N-C_6H_4-CONHC_6H_{11}$

(96)

$C_6H_{11}CONH-C_6H_4-N{=}N-C_6H_4-NHCOC_6H_{11}$

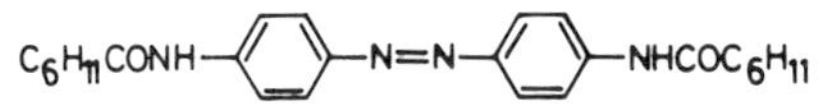

(97)

FIGURE 5.3 Poly azoaromatic amides (94–95) and model compounds (96–97).

ergy of activation should be substantially higher for hindered rotation in polymer backbones than for a similar process in a small molecule resulting in a large decrease in the rate of conformational transition.[17] The low value of *trans-cis* energy barrier of azobenzene would make the study of photoisomerization of azobenzene residues in the backbone of polymers and their analogs eminently suitable for the characterization of the general behavior of conformational transitions involving low energy barriers.

The photoisomerization of polymeric *trans* azo compounds was initially studied by Morawetz and coworkers[18-20] who first demonstrated what effect on azo linkage incorporated into the chain would have as compared to their corresponding low-molecular weight analogs. Two series of modified nylon-6,6 polyamides were synthesized (Figure 5.3) by the Morawetz group. The isomerization around the double bond was followed kinetically by UV spectroscopy such that any dispersion of rate constants could be characterized. Irradiation of the *trans* polymers (94) and (95) in formic acid solution produced photostationary states comprised of mixtures of polymers differing only in *trans-cis* linkages. Morawetz and Tabak[18] studied the dark reaction and found no difference between the rates of isomerization of azobenzene residues in the chain backbone when compared with their low molecular weight analogs (96,97). However, this data provided no conclusive evidence against the crankshaft like motion, since the chain contained many bonds with lower energy barriers. Chen and Morawetz[20] studied the rate of photoisomerization, where the energy barrier is low, and found that at high dilution the polymer and the low molecular weight analog behaved in a similar manner (Figure 5.4). However, when the photoisomerization of the low molecular weight compound was followed in solutions containing added polymer, the rate of its isomerization changed very little, while the photoisomerization rate of azobenzene polymers had their rate reduced by many orders of magnitude, because of the cooperativity of conformational changes in the polymer chains. Thus the "crankshaft-like" motion has now been disproved both by computer simulation[21] and experimentally.[22]

Kumar et al.[23,24] reported that the introduction of azobenzene groups in the main chain of polyureas does not impede isomerization in dilute solution. A series of azo polyaromatic ureas were synthesized incorporating pyridine moieties as polymeric ligands. The activation energies for *cis-trans* thermal isomerization of these polyureas suggest similar energy barriers to low molecular weight analogs as well as to polymers with different molecular weights and chain segments. Lamarre et al.[25] reported that, in dilute solution, the ability of the azo moiety to isomerize photochemically is not strongly influenced by the composition of the polyurethane chain, since the choice of a particular diol used for the polymer does not make much difference to the *A* values. These observations are not surprising in view of the much faster scale of photoisomerization in comparison to the motions of the polymer chain as demonstrated by Irie and Schnabel.[26] By using the flash photolysis technique, they reported that the *cis-trans* photoisomerization of the azobenzene moiety in an aromatic polyamide in solution occurs to about 90% during a 20-ns flash, with the remaining 10% complete within 100 ns after the flash. They also measured the relaxation time due to the conformational change of the total molecule in dilute solution following photoisomerization, using time-

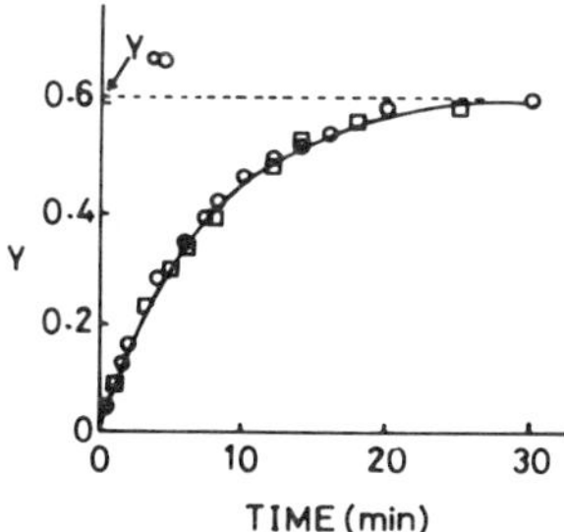

FIGURE 5.4 Course of photoisomerization of (94) and (96) in dilute solution. Y_∞ fraction of *cis*-azobenzene residues at photostationary state. Reprinted in large part with permission from *Chem. Rev.* 1989, 89:1915–1924. Copyright 1989 American Chemical Society.

resolved light scattering measurement, and found it to be about 1 ms—much slower than the photo chemical reaction.

B. Side Chains

The physical property of a polymer can also be altered by an induced conformational change in the side chains of a polymer. Copolymers containing an azo group in the side chain have been investigated by Komogawa et al.[27-30] who prepared several model azo compounds for comparison with the corresponding appended polymer species. The azo bearing polymers (98) and their models (99) were compared in solution as well as in film form (Figure 5.5). By measuring the recovery half-time periods they found that the reaction was slowest with comonomers which increase the rigidity of the chain, e.g., styrene or MMA compared to methyl acrylate. As one would expect, the recovery times were five times longer in polymer films than in solution. This was attributed to the increased steric interference by the polymer upon the *cis* to *trans* isomerization of azo compounds. One can propose that the photoisomerization step from *trans* to *cis* induces local strain or hindrance which causes the increased rate of thermal return. This was particularly noticeable in two cases where the azo group was appended to the polymer backbone in a manner causing steric interference. These investigations infer that photochemically induced strain assisted the thermal return in both systems where the azo groups are appended in polymer matrices.

Studies of polymers in bulk were pioneered by Morawetz and Paik[19] who found that the dark reaction of azobenzenes fixed as polymer side chains had the same rate of return both in rubbery polymers and dilute solutions. In the glassy state, an anomalous fast component was observed,

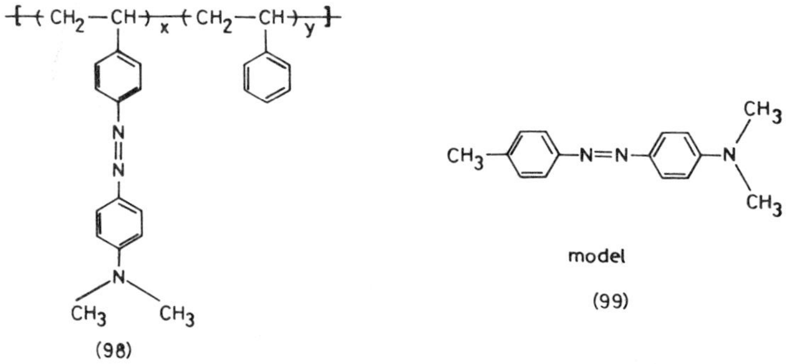

FIGURE 5.5 Structures of styrene copolymer with pendant dimethyl amino azobenzene(98) and corresponding model compound (99).

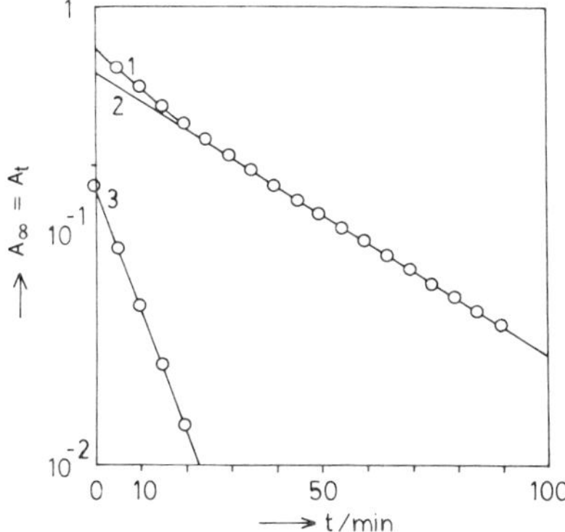

FIGURE 5.6 Resolution of thermal *cis-trans* isomerization of the azo chromophore in copolymer of ethyl methacrylate at 70°C using two first order reactions. A_∞, A_t absorbance at $\lambda = 343$ nm at infinite time and at time t (1) experimental curve, (2) normal isomerization curve, (3) fast reaction curve. Reprinted in large part with permission from *Chem. Rev.* 1989, 89:1915–1924. Copyright 1989 American Chemical Society.

an effect reported previously by Priest and Sifain[31] for small azoaromatics dissolved in glassy polymers. Priest and Sifain interpreted this as evidence for the nonuniform distribution of free volume in the nonequilibrium glassy state. Important work on the volume relaxation of glassy polymers as studied by means of the kinetics of photochemical isomerization was reported by Sung.[32,33] In the system studied, a fraction of the azobenzene residues isomerizes very slowly and that fraction is a function of the distance from the glass transition temperature.[25] The magnitude of T_g depends on the way the azobenzene is attached to the polymer.[32] In a brilliant recent paper,[33] the time of dependence of T_g for azobenzenes attached to polystyrene in different ways is interpreted in terms of the changes in free volume distribution during volume relaxation.

More recently, a similar photochromic behavior was observed by Eisenbach[34-37] for the *cis-trans* isomerization of azobenzene residues attached as side groups to copolymers when examined below their T_g. These copolymers were obtained by the copolymerization of the monomers 4-methacryloyl amino azobenzene and various acrylates and methacrylates. In solution and in rubbery state the thermal recovery follows first order kinetics; (Figure 5.6) in the glassy state, however, some azo groups react anomalously fast, others isomerize much slower nearly as in solution. These data confirmed earlier observations of Paik et al.[19] in the case of MMA-STY copolymers having azobenzene side groups. Here also the decoloration curves can be resolved by two simultaneous first-order reactions. Most striking, however, is the dependence of the rate of bleaching in the neighborhood of the T_g of the copolymer. The change of E_a and activation entropy around T_g was interpreted on the basis of physical changes of the polymeric environment, i.e., an important increase in the chain segment mobility and consequently of the free volume above T_g.

The temperature dependence of melt viscosity at temperatures considerably above T_g approximates an exponential function of the Arrhenius type. However, near T_g the viscosity-temperature relationship for many polymers is in better agreement with the Williams-Landel-Ferry (WLF) treatment.[38] These authors showed that with a proper choice of reference temperature, T_g, the ratio of the viscosity to the viscosity at the reference temperature could be expressed by a single universal equation—the WLF equation.

Eisenbach[34-37] also showed the general validity of the WLF equation in the interpretation of the photochromism of azobenzenes and spiropyrans linked to amorphous bulk copolymers. Eisenbach used directly the WLF equation:

$$\log a_t = -C_1 (T - T_g)/(C_2 + T - T_g)$$

where a_t is the ratio of the relaxation time at T to that of T_g ($a_t = KT_g/KT$). On the basis of this equation, he determined graphically the C_1 and C_2 for several photochromic systems. The corresponding numerical values were nearly constant for a group of polymers, e.g., for photochromophores dissolved in different polyacrylates and polymethacrylates and attached to the latter as side groups. They differ, however, considerably below and above the T_g of the polymer system. It should be noticed that these experimental C_1 and C_2 values differ markedly from the universal WLF constants, likely because the photochromic behavior is mainly controlled by the free volume distribution around the photochrome, which is different from its usual random distribution.

C. *Crosslinks*

The local orientation of rubber-like materials is of considerable interest for the understanding of the overall mechanical behavior of these materials. The experimental results were mainly interpreted by a greater degree of orientation of the chain segments near crosslink junctions. This concept can be rationalized taking into account that near a crosslink junction the restrictions acting on a chain segment mainly arise from branches while only two chain branches act on segments in a chain (Figure 5.7). These conformational restrictions, as a consequence, should be a function of the crosslink density. A final test of this concept involves direct measurement of the restrictions. One possibility of doing this is to introduce conformational changes by a chemical reaction and to detect the response of the elastomeric network to this disturbance.

Kumar et al.[39] investigated the question of how the photochromism of azo groups lodged in the crosslinks would be influenced by the nature of the polymer matrix and by the presence or absence of spacer groups separating it from the main chain. As found in linear or branched polymers, photochemical *trans-cis* isomerization of the azobenzene units in the crosslinked polymers can be achieved by irradiating at 360 nm. A series of crosslinking agents (100,101,102) containing azobenzene units with amide and sulfonamide spacer groups were prepared (Figure 5.8). UV irradiation creates *cis* isomer in high yields in each case, except in the case of sample (102). A bulky substituent next to the azobenzene seems to retard the photoisomerization process.

From the comparison of the first-order plots of (Figure 5.8) thermal isomerization rates, it was evident that the back reaction is not affected so much by the nature of the polymer matrix, but it is affected by the way the chromophore is fixed. The rate of thermal recovery is fastest in the case of the sulfonamide spacer group having an adjacent phenyl group. These investigations point out the importance of structural and steric environment surrounding the crosslinks.

Stadler et al.[40,41] prepared swollen gels of polyisobutylene crosslinked with (103). The butyl rubber gels are highly swollen and thus the isomerization behavior should be similar to that of a dissolved polymer. The reaction rate constant depends on the crosslink density. From measurements at various temperatures it is deduced that the activation enthalpy remains unaltered. The observed dependence of the reaction rate constant is caused by the dependence of the activation entropy on crosslink density. The dependence of the isomerization rate on the crosslink density can be interpreted in terms of the classical theory of rubber elasticity (Figure 5.9). The isomerization causes a decrease in the conformational entropy of

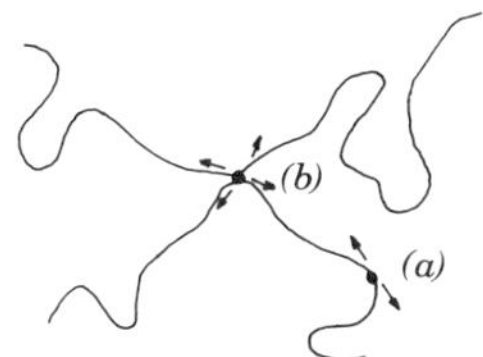

FIGURE 5.7 Schematic representation of the restrictions acting on a polymer segment in a chain (a) and near crosslink junction (b).

$CH_2{=}CH{-}C_6H_4{-}N{=}N{-}C_6H_4{-}CH{=}CH_2$

(100)

$CH_2{=}C(CH_3){-}C({=}O){-}HN{-}C_6H_4{-}N{=}N{-}C_6H_4{-}NH{-}C({=}O){-}C(CH_3){=}CH_2$

(101)

$C_6H_5{-}CH{=}CH{-}SO_2{-}NH{-}C_6H_4{-}N{=}N{-}C_6H_4{-}NH{-}SO_2{-}CH{=}CH{-}C_6H_5$

(102)

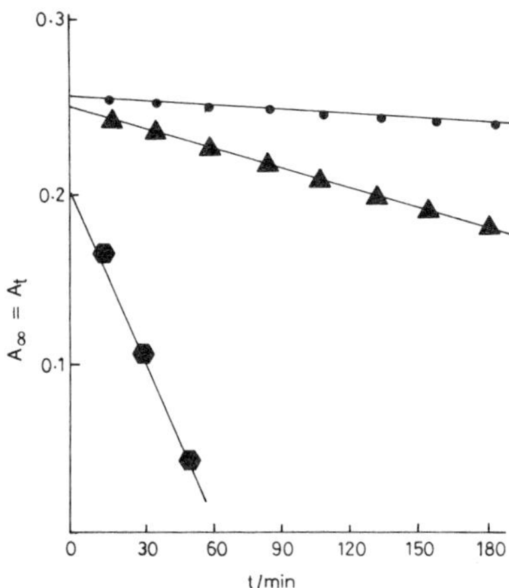

FIGURE 5.8 First-order plots for the thermal *cis-trans* isomerization of the azo chromophore with different spacers in crosslinks of polystyrene (A_∞ and A_t are absorbances at 350 nm at infinite time and time t); temperature 30°C; solvent–CH_2Cl_2; (●) crosslinked with **(100)**; (▲) with **(101)** and (⬢) with **(102)**.

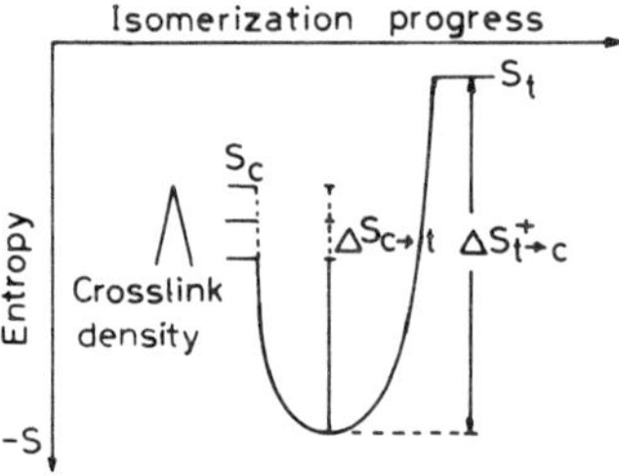

FIGURE 5.9 Schematic representation of the entropy changes for the isomerization process in the networks [Sadler et al., *Polymer* 27:1254 (1986)]. Reprinted in large part with permission from *Chem. Rev.* 1989, 89:1915–1924. Copyright 1989 American Chemical Society.

the chains. The corresponding elastic retractive forces induce a faster *cis-trans* isomerization. These results seem to be in contrast to those observed by Eisenbach[35] where it was deduced from kinetic measurements that obviously crosslinking itself and also the degree of crosslinking retards the thermal isomerization process. However, a more detailed analysis of the data shows that this dependence reflects the different T_gs of the systems investigated, so that the observed rate constants depend primarily on the viscous properties of the surrounding polymer matrix and can be interpreted in terms of the effective free volume concept, where generally the differences in the activation enthalpy are considered. In these studies, the effect of the surrounding matrix diminishes because the T_g of the gels is far below the temperature of the measurement. Owing to the high mobility of the swollen state the chains immediately reach their conformation corresponding to the thermal equilibrium. As a consequence, the activation enthalpies are constant which is not the case for polyacrylate networks.

Sung et al.[42,43] reported a technique for characterizing the cure of epoxy by the azo chromophore labelling technique. A small amount of p p′-diamino azobenzene (DAA) was used as a reactive label in a model epoxy consisting of diglycidyl ether of bisphenol A and diaminodiphenylsulfone (DDS). The reactivities of DDS and DAA are similar and the cure process is followed by the UV-VIS spectral changes of DAA occurring above 400 nm. As the epoxy is cured, the max of the $\pi - \pi^*$ transition corresponding to the azo bond of DAA shows red shifts allowing the spectral discrimination for four major cure products, namely crosslinks, branch points, linear chains and chain ends. Deconvolution of the UV-VIS spectra based on the band assignments of the model compounds representing cure products provides a quantitative estimate of the four cure products as a function of temperature or cure time. DAA-labelled epoxy (DGEBA-DDS) also exhibits very sensitive changes in fluorescence intensity corresponding to the emission by the DAA label as the function of cure extent.

5.3.2. Photoresponsive Polymers: Azobenzene as a "Trigger"

Photoresponsive synthetic polymer is a speciality polymer having photoreceptor chromophores, which can transfer light energy into a change in the conformation of the polymer.[44] The light is once stored in a chemical structure change of the chromophore and then transferred into the polymer chain, causing reversible conformational changes. The conformational change should produce a concomitant change in physical and chemical properties of the polymer solutions and solids. This is the basic idea for controlling properties of polymers by photoirradiation using the photoresponsive trigger molecules. Several examples of photostimulated physical property changes of photoresponsive polymers having azobenzene residues are described. Some of the physical and chemical properties which are reversibly controlled include (a) viscosity, (b) conductivity, (c) pH, (d) solubility, (e) wettability, and (f) mechanical property.

A. Photoviscosity Effects

The viscosity of a polymer system is a rather direct reflection of the polymer conformation. The constitution of azo aromatic polymers containing azobenzene units in the backbone suggests that these polymers would behave like semi-flexible rods in solution. The extended rod-like shape of the semi-flexible chain is expected to shrink rapidly to a compact conformation when the configuration of the constituent azobenzene units changes from the *trans* to the *cis* form. It results in the change in viscosity of the polymer solutions and is termed as the photoviscosity effect.[45]

Irie et al.[46,47] synthesized the following polyamides in an attempt to

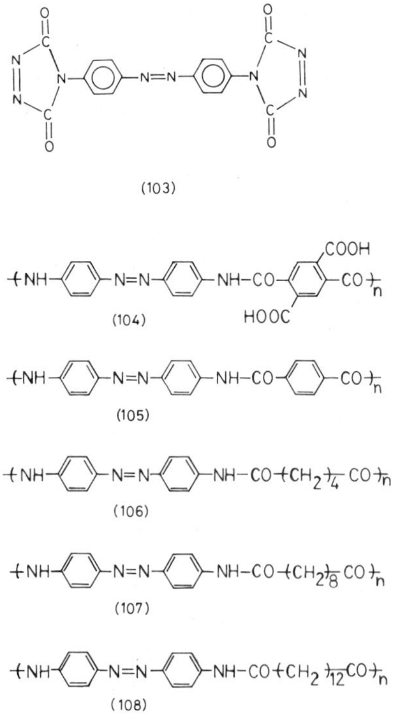

(103)

(104)

(105)

(106)

(107)

(108)

FIGURE 5.10 Photoresponsive polyamides.

control solution viscosity of polymers(**104–108**). Intrinsic viscosity of polymer (**104**) in N,N-dimethyl acetamide decreased from 1.22 to 0.5 on UV irradiation (410 > λ > 350 nm) and returned to the initial value in 30 hrs in the dark at 20°C. The slow recovery of the viscosity in the dark was accelerated by visible light (λ > 470 nm). On alternate irradiation of UV and visible light, the viscosity reversibly changed as much as 60%. The similarity of the response of the viscosity and the *trans* form content to photoirradiation suggests that the photodecrease and increase of the viscosity directly correlated with the isomerization of the azobenzene residues in the backbone.

When the azobenzene units are connected by rigid phenyl groups, the resulting viscosity change should be large, while it should become small, when the connecting groups are flexible like long methylene units. Irie et al.[46] demonstrated the absence of a photoviscosity effect when the methylene chain is long enough to act as a strain absorber.

Similar experiments were carried out by Blair et al.[48] for polyamides having azobenzene residues in the backbone of polymers. They could not observe any decrease in the viscosities under UV action although a small decrease was detected in the reduced viscosities at high concentrations. The absence of photodecrease is possibly due to the inclusion of flexible piperazine segments in the polymer chains.

Kumar et al.[23,24] reported the photodecrease in viscosity for a new class of azoaromatic polyureas. Since there is no effect of concentration on the viscosity of the sample, this photodecrease must arise from a conformational change of the polymer chain and not from interchain interaction. Thermal recovery of the original viscosity as well as the recovery of the absorption intensity is compared in Figure 5.11. These two plots superimpose well and indicate that the contraction of the macromolecular volume is indeed induced by the isomerization of the azobenzene residues and that slow thermal recovery expands the chain conformation.

Lovrien and Waddington,[49] in their pioneering work, designed two systems for controlling the polymer conformation by light. In the first, a polyelectrolyte bears covalently bound side chains capable of undergoing photochromic *trans-cis* isomerization. The net charge on the polyelectrolyte tends to extend the polymer via electrostatic forces, and the *trans* forms of the side chains generate nonpolar interaction forces that tend to contract the polymer. Illumination generates *cis* forms, changing these hydrophobic interactions somewhat and causing the polymer to reach new equilibrium by expansion of the coil. Photoresponsive materials of this first type were prepared by copolymerizing acrylic or methacrylic acid with 4-acrylamido azobenzene or its 2,2 dimethoxy derivative. Viscosity increase was expected and found in these systems upon irradiation.

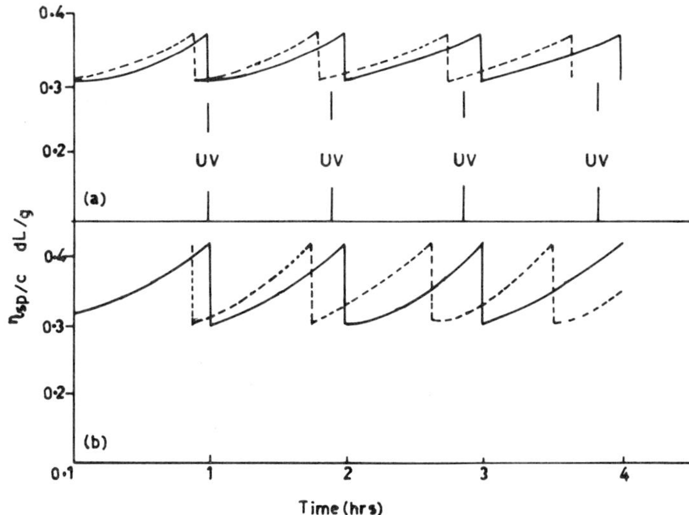

FIGURE 5.11 Changes of (a) *trans* form of azobenzene residues in polyureas derived from diaminopyridine, MDI and diamino azobenzene (PU–MDU–DAP), (b) viscosities of the same polymer solution upon alternate irradiation at 35°C (——) in the dark, (- - -) in visible light.

B. Photomechanical Effects

To what extent conformational changes of the chromophores and connected chain segments of a photochromic polymer may induce changes of dimensions of bulk polymers and thus generate reversible photomechanical effects is an area of great interest. From the organic point of view, reversible dilation/contraction phenomenon should be observed in photochromic networks above T_g, i.e., in the rubbery state where segment mobility is important. Considering that the isomerization occurs in the rubbery state at a rate similar to that in solution, it should be expected that the use of isomerization reactions for photocontractile behavior should be optimal with easily deformable networks.

Photoisomerizable groups incorporated into the polymer framework have been shown to cause reversible contraction or expansion of polymer samples upon irradiation. Agolini and Gay[50] prepared a polyimide(110) containing azoaromatic groups in the backbone as shown below. The final imide polymer was obtained in a semicrystalline film containing the azo group predominantly in the *trans* form. These films contracted on heating and irradiation, probably due in part to the isomerization from *trans* to *cis* form in the amorphous regions of the film. However, the rates of contraction and dilation were not controlled linearly by the rate of isomerization. Stress was produced slowly when a sample held to constant length was irradiated. In the dark at lower temperatures the films returned to the unstressed form. Similarly reversible contraction/dilation experiments under constant load were recently performed by Eisenbach[51] on stretched poly (ethyl acrylate) networks, crosslinked with 4,4 dimethacryloyl amino azobenzene. Upon irradiation *trans-cis* isomerization caused conformational changes of adjacent network segments which are considered to be responsible for the photomechanical effect. The observed contraction amounted to only about 0.15–0.25%.

Blair, Pogue and Riordan[48] described photoresponsive effects of photochromic polyamides in which every monomer unit contains an azo group, namely the 3,3′-azodibenzoyl, trans-3,5 dimethylpiperazine(111) and its 4,4 isomer(112). On irradiation the measured stress increases indicating a contraction of the sample up to a photostationary state; in the dark the stress decreases again, and the cycle can be repeated many times. Here again the relaxation time in the dark is much more rapid than the normal *cis-trans* in the solution and should be due to the isomerization of a small fraction of azo residues on the local strained nonequilibrium sites in the polymer matrix.

Blair et al.[48] have also carried out Langmuir film balance measurements on monolayers of the photochromic polyamides mentioned above. Changing from dark to light reduction in area per monomer unit was observed and interpreted as resulting from azo *trans* to *cis* isomerization. The difference between the pressure area curves of the meta- and para-

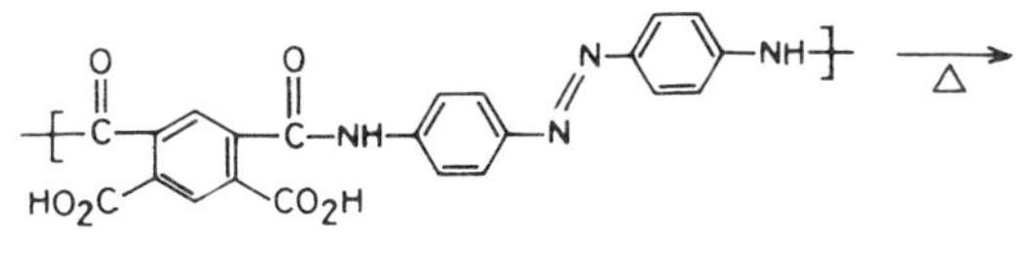

(109)

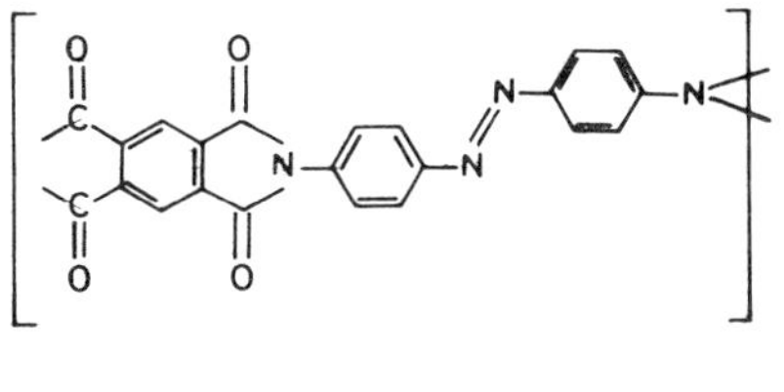

(110)

FIGURE 5.12 Azoaromatic polyimide.

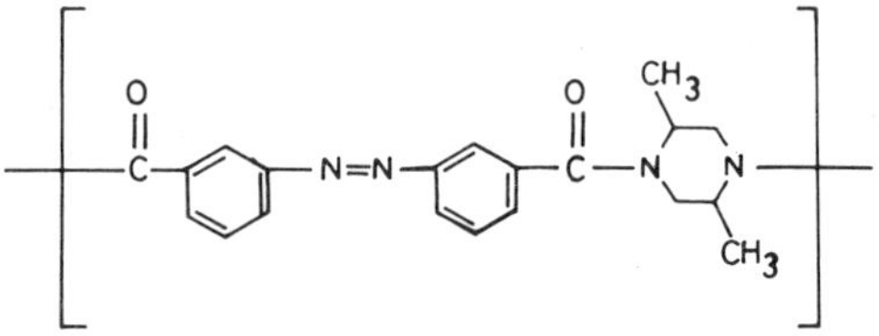

(111)

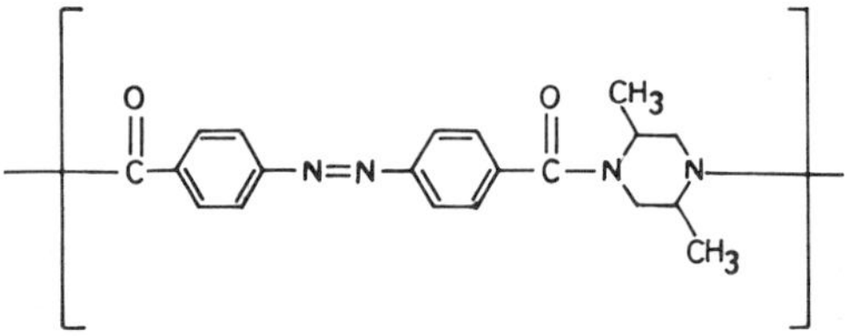

(112)

FIGURE 5.13 Azoaromatic polyimides which exhibit a photomechanical effect.

polyamides were interpreted on the assumption that the para compound is linear and the meta compound is helical.

Besides the photomechanical effects observed on rubbery networks and swollen gels, photocontractibility of the photochromic systems in the solid state without the chemical crosslinking agent has been described. In some cases possibly partial crystallinity and eventually hydrogen bonding ensure physical crosslinking and consequently a network structure. In most simple cases, the photosensitive compounds were dissolved in a polymeric matrix and the effects are based on the interactions between the dye and the polymer substrate molecule.

C. Reversible Solubility

When photoisomerizable chromophores are incorporated into the backbone of the polymer chain or as pendant groups, photoisomerization of the chromophores will affect the physical properties of the polymers and the polymer solutions, especially if the isomerization involves a change of polarity. Changes of dipole moments of pendant groups upon photoirradiation alter the balance of interchain interaction resulting in expansion/contraction of the polymer chain. Irie et al.[52] reported the reversible solubility change of polystyrene (PS) in cyclohexane. PS with a small amount of azobenzene pendant groups (5 mole% of monomer unit) became insoluble in cyclohexane upon irradiation with UV light, while low molecular weight azobenzene itself did not show any solubility change upon photoirradiation. On visible irradiation, the polymer again became soluble. The dynamics of the formation of the precipitation in cyclohexane was studied by a laser photolysis measurement combined with a light scattering detection method. The precipitation of the polymer upon irradiation with UV light was followed by measuring transmittance at 650 nm, where azobenzene has no absorption. In cyclohexane, intermolecular interaction between polystyrene and the solvents is in balance with the intra- and interpolymer interactions. The dipolemoment increase of the pendant groups by UV irradiation is considered to alter the balance of polymer-solvent and polymer-polymer interactions. The introduction of the non-polar *trans* form of azobenzene into polystyrene pendant groups scarcely affects the polymer-solvent interaction in cyclohexane while the polar *cis* form tends to decrease the polymer-solvent interaction. Therefore, upon UV irradiation, the polymer-solvent interaction decreases until the polymer precipitates. The amount of precipitation sharply increased when the azobenzene content exceeds 5 mole% of monomer units. The content indicates that isomerization of a few mole% of azobenzene units in the polymer chain is enough to cause a solubility change of the polymer.

Irie et al.[53] also reported a reversible change of the phase separation temperature of aqueous polymer solutions by photoirradiation. Photocontrol of the phase separation of aqueous solutions of poly(N-isopropyl acrylamide) by introducing azobenzene moieties into the pendant groups was achieved by copolymerization of N-isopropyl acrylamide with N-[4-(phenyl azo) phenyl] acrylamide.

D. Reversible Surface Free Energy: An Interfacial Approach

The surface free energy of a solid is an important feature of printing, dyeing and adhesion.[54] If surface free energy which is the inherent value of the material can be controlled by external physical signals such as light, wide application of new materials are expected. Ishihara et al.[55-57] have synthesized photoresponsive polymers that contain a photochromic azobenzene group in their side chains. The contact angle formed by water on the surface of the film prepared from p-phenyl azo acrylanilide-2-hydroxyethyl methacrylate copolymer is changed by the photoisomerization of the azobenzene moiety. Further, photoinduced changes in the surface free energy of the azo aromatic polymer, prepared by the introduction of the azobenzene groups into the side chains of hydrophilic poly(HEMA) was investigated.

Figure 5.14 shows the change of the wettability of the polymer surface by water and the absorbance at 325 nm, which corresponds to a peak for

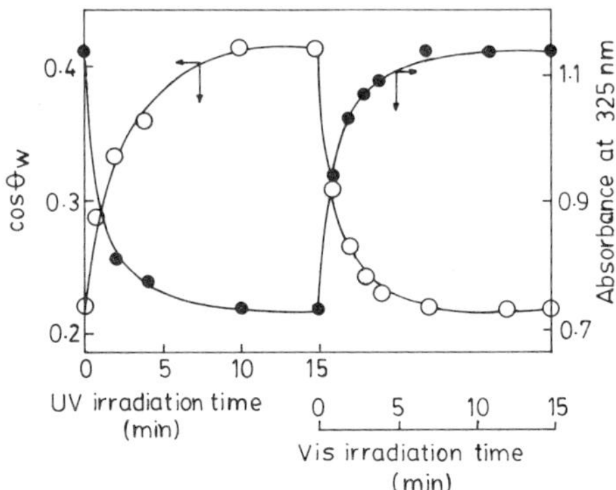

FIGURE 5.14 Photoinduced change in wettability and absorbance of azoaromatic polymer film: θ_ω contact angle by water; (– ● –) UV light, = 350 nm: (– ○ –) visible light < 470 nm. Reprinted in large part with permission from *Chem. Rev.* 1989, 89:1915–1924. Copyright 1989 American Chemical Society.

trans-azobenzene moiety, when the light was irradiated onto a film of azoaromatic polymer. When the irradiation is carried out, the absorbance of the *trans* form decreased with irradiation time and the photoequilibrium was obtained within 15 min. The absorbance change was accompanied by an increase in the wettability of the surface of the polymer film. If irradiation by visible light follows, the absorbance of the *trans* form returned to its original level and the wettability decreased once again. The results indicate that the wettability of the polymer surface can be regulated by the photoisomerization of the azobenzene moiety. Irie et al.[58] also reported that the contact angle of water on butyl methacrylate-2-(hydroxyphenyl)-4-(vinylphenyl) benzyl alcohol copolymer surfaces increased greatly upon UV irradiation and reversed in the dark.

An investigation was made on the adsorption behavior of water-soluble azo dye toward a polymeric adsorbent, and it was demonstrated that a reduction in the hydrophobic interaction brought on by photoisomerization greatly contributes to the desorption of azo dyes from the adsorbent.[55] Water-soluble polymeric azo dyes were prepared and the photoinduced adsorption desorption behavior of the surfactant was examined.

Ishihara et al.[59] also reported the regulation of the hydrophobic chromatography for proteins by light. Polymeric hydrophilic adsorbents containing an azobenzene moiety as a ligand were prepared and the photoinduced adsorption-desorption behavior of proteins was investigated. The separation of protein mixtures was also investigated using a gradient column which was constituted by two polymeric hydrophilic adsorbents having different hydrophobicities.

E. Photochromic Polypeptides: Helix Reversal

Visual purple rhodopsin is composed of retinal and a protein called opsin. While rhodopsin absorbs light energy, its retinal component undergoes photoisomerization accompanied by a change in conformation of opsin.[60] This light-induced conformational change is a trigger which excites the nerve cells of the retinal rods. Recent progress in polypeptide chemistry has prompted several workers to construct systems analogous to that in the visual excitation which permits light-induced conformational changes of polypeptides.

The idea of searching conformational changes induced by light on modified polypeptides was originally due to Goodman et al.[61,62] but it was only later that conformational changes upon photoisomerization of the side chain chromophores were claimed to be observed. The photoinduced conformational changes of the polypeptides were first accomplished by Ueno et al. and Ciardelli et al.[63-69] with azobenzene-containing poly(aspartates) in solution (Figure 5.15). They have found that photoresponsive poly(aspartates) can undergo a photoinduced conversion of the secondary structure in solution. However, they also reported that the backbone conformation of azobenzene-containing poly(aspartates), which undergo a

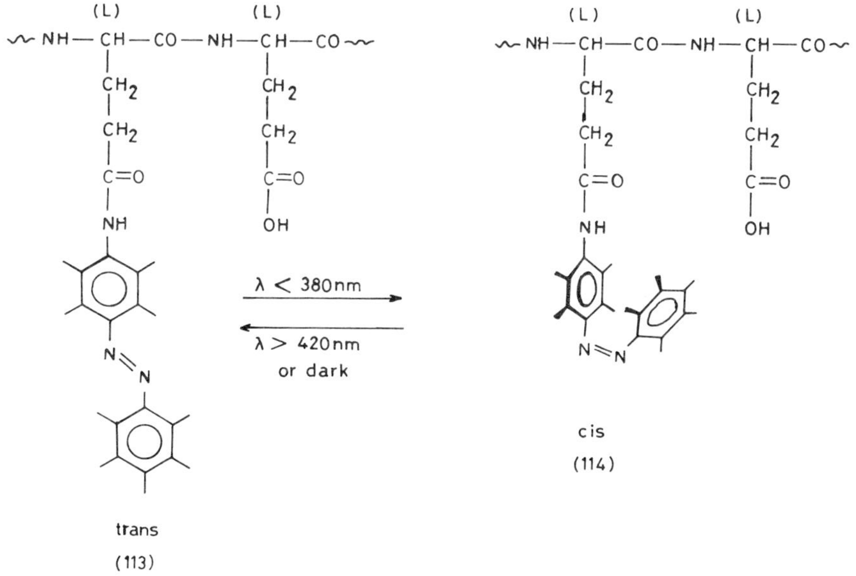

λ < 380nm

λ > 420nm or dark

trans (113)

cis (114)

FIGURE 5.15 *Cis-trans* isomerism of azo chromophore bound to poly (L-glutamic acid).

photoinduced conformational transition in solution, could not be changed by irradiation in the solid membranes.[70] Ciardelli et al.[71] studied the water-soluble azo modified PGAs in an aqueous solution and reported photoinduced α-helix coil and β-structure-coil transitions that depend on the azo group content and pH value of the aqueous solution at which irradiation is carried out. The CD spectra of the azo modified poly(glutamic acid) solid membrane containing 14 mole% azo groups, which are adapted in the dark or irradiated in water (pH 6.2), revealed that the UV irradiation does not induce any changes of the backbone conformation of the membrane. Later work was concerned with photoisomerization of the azo moieties in polymers of phenyl alanine, aspartic acid and L-lysine.[72-74]

F. Reversible pH Change

Light-induced pH change is of particular interest in connection with the function of bacteriorhodopsin. Bacteriorhodopsin, found in the purple membrane of *Halobacteria*, uses light energy to translocate protons across the membrane. The cell uses the energy stored in the electrochemical gradient for ATP synthesis.[75]

Photocontrol of the conformation of polymethacrylic acid in solution at degrees of ionization of 0.2 (pH 4.5–5.5), at which the polyelectrolyte molecule uncoils, is possible with the use of cationic ligands based on p-phenyl azophenyl trimethyl ammonium iodide or the copolymers.[76] Solutions of polymethacrylic acid with cationic ligands exhibit a decrease of 0.2 units in the pK_a on *trans-cis* photoisomerization, caused by a change in the degree of coiling of the polyelectrolyte molecules. The photochromic copolymers also display an induced photochemical jump in pK_a when the azo groups pass into the *cis* form. Since

$$pK = pH + \ell n[(1 + \alpha)/\alpha]$$

the degree of ionization of polymethacrylic acid modified by photochromic groups, and hence the conformation of the polyelectrolyte, should be controlled photochemically in buffer solutions.

Irie et al.[52] reported that electrical conductivity of N,N-dimethyl acetamide solution containing a polymer (**104**) exhibited a response on alternate irradiation of UV and visible light. The response of the conductivity correlated well with the isomerization of the azobenzene residues in the backbone. The correlation suggests that dissociation equilibrium of amide substituted terephthalic acid residues in the polymer backbone is influenced by conformational change of the polymer chain. Dissociation of the acid is stimulated in the compact or shrunken conformation, while the extended conformation depressed the dissociation.

G. Photoresponsive Metal Ion Chelation

Host molecules have drawn considerable attention as simplified enzyme model systems. It is well known that the specificity is based on the spatial "fitness" between host and guest molecules. It is therefore expected that if the conformation of the host molecules is somewhat distorted by the change in the photoinduced "trigger," it would lead to the change in the binding ability. If it really occurs, one can control the functions of the host molecules by light. With these objects in view, Shinkai et al.[77-80] combined crown ethers as a functional group with azobenzene as a photoantenna and have attempted the photocontrol of the crown ether functions. They reported that these photoresponsive phenomena are applicable to the photocontrol of the binding properties of crown ethers on polymer (Figure 5.16). Earlier, it has been established that *cis* crown ethers and polymeric crown ethers favorably bind alkali metal ions with large ionic radii through the formation of intramolecular 1:2 metal/crown sandwich complexes. Shinkai et al. synthesized a polystyrene derivative bearing the pendant 4-azo (benzo-15crown-5)(**115**) and its monomeric model (**116**) and estimated the photoresponsive affinity towards alkali metal ions (Figure 5.16). The study demonstrated that the ion binding ability of the crown ether immobilized in the polymer changes in response to the changes in the conformation and side chain configuration.

Kumar et al.[23,24] prepared a series of azoaromatic polyureas with incorporated pyridine and bipyridines as polymeric ligands. The activation energies for *cis-trans* thermal isomerization of these polyureas suggest similar energy barriers to low molecular weight analogs. An increase in E_a

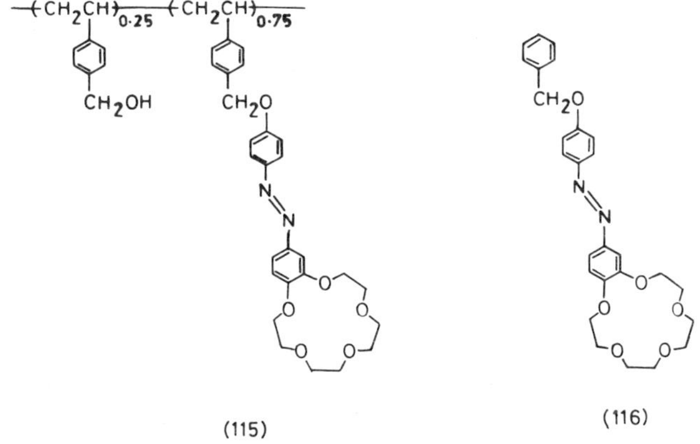

FIGURE 5.16 Photoresponsive crown ethers.

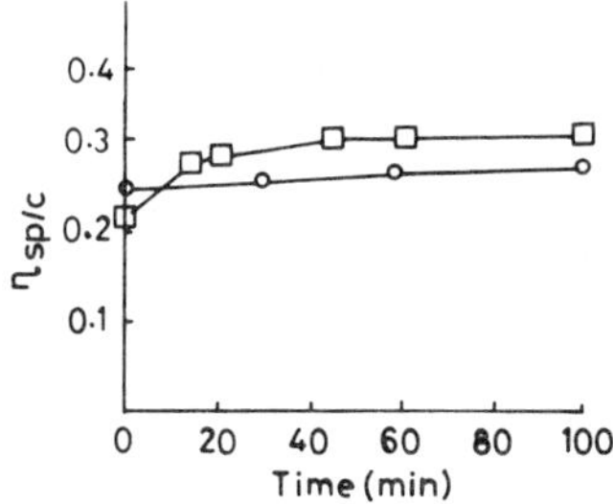

FIGURE 5.17 Changes in viscosity of polyurea of diamino pyridine, 4,4′ aminoazobenzene and MDI (1:1:2) solution with progressive complexation of $Co(OAc)_2$ in the dark (○); when the solution is preirradiated with UV (□). Concentration of polymer is 0.8123g/dL. $[Co(OAc)_2] = 1 \times 10^{-4}$ mol. Reprinted in large part with permission from *Chem. Rev.* 1989, 89:1915–1924. Copyright 1989 American Chemical Society.

values was observed upon complexation with cobalt and nickel. The viscosity of the solution increases after complexation. Preirradiated polymer solutions showed a greater increase in viscosity and slightly larger incorporation of metal ions compared to its behavior in the dark. This observation is attributed to conformational rearrangement of the macromolecular ligand prior to complexation (Figure 5.17).

H. Membrane Permeability

Much effort has been made in the characterization of the photoresponsive properties of the membranes entrapping photochromic compounds such as azobenzene and spiropyran derivatives. Anzai et al.[81,82] used the *cis-trans* isomerization of azobenzene modified crown ethers to regulate the membrane potential across a polyvinyl chloride membrane. Photoinduced *trans-cis* isomerization of azobenzene derivatives has also been widely used to photocontrol metal ion permeabilities through PVC membranes,[83,84] liquid crystalline membranes,[85] liposome[86] and bilayer membranes.[87] More recently, it has been shown that the membrane potentials and permeabilities of polymer membranes containing photochromic compounds in the polymer chains could be regulated by light irradiation.

Morgan et al.[88] synthesized a phospholipid molecule bearing an azobenzene linkage within one acyl chain. The lipid has been incorporated into vesicles of dipalmitoylphosphatidylcholine, (DPPC) and the effects of photoisomerization on the vesicles permeability and phase behavior studied by light scattering and fluorescence spectroscopy. The phase transition temperature of DPPC is reduced, and the transition is broadened by the *trans* azo lipid. After photolysis the transition temperature is further reduced and non-equilibrium effects are evident. Vesicles containing the azolipid can sustain a pH gradient before photolysis but pH slowly equilibrates following irradiation. Results indicate that photoisomerization causes no loss of bilayer integrity.

5.4. Conclusion

Photoresponsive polymers described in this chapter represent a new field of speciality polymers. As frequently happens, a field that is essentially derivative develops a *raison d'etre* of its own. Thus, the photomechanical and photoviscosity effects, pH control systems, are important by themselves and have possible commercial applications. Although they have not yet been used practically, it is possible for them to find applications in constructing photoactive devices in several fields in the near future, such as printing, photocopying, photolithography, photosensory and so on.

An example of such applications, which attract current attention, is speciality organic materials which can store memories in the molecular levels. If one could succeed in designing a molecule which shows a reversible response in its physical properties to semiconductor lasers in infrared regions, one would find certainly many applications in electronics as information memory devices. Another possibility is to use photomechanical effects. Recently, many interests have been directed to shape memory metals, which memorize the shape and restore the shape after heated to the critical temperature. Photoresponsive polymers could be used in similar fields as the shape memory metals aim at, such as photomechanical engines or photoswitches.

Although the photochemistry of azobenzene-containing polymers has grown vigorously, very few other azoaromatic chromophores have been used as photochromic units. I believe this to be a clear disadvantage in widening the applications of these polymers. Studies with other azoaromatic chromophores might eliminate some of the kinetic problems which these polymers seem to have in terms of practical utility. Almost total lack of interest in designing photolabile polymers based on aliphatic units may be understood in terms of low quantum yields of photolysis expected because of photoisomerizations.

5.5. References

1. Engel, P. S. 1980. "Mechanism of the Thermal and Photochemical Decomposition of Azoalkanes," *Chem. Rev.*, 80:99.

2. Zollinger, H. 1961. *Azo and Diazochemistry*, New York, NY: Interscience.
3. Griffiths, J. 1971. "Photochemistry of Azobenzene and Its Derivatives," *Chem. Soc. Rev.*, p. 481.
4. Zimmerman, G., L. Y. Chow and U. I. Paik. 1958. "Isomerization of Azobenzene," *J. Am. Chem. Soc.*, 80:3528.
5. Gordon, M. S. and A. Fischer. 1968. "A Molecule Orbital Study of the Isomerization: Mechanism of Diazocumulene," *J. Am. Chem. Soc.*, 90:2471.
6. Nerbonne, J. M. and R. G. Weiss. 1978. "Elucidation of Thermal Isomerization Mechanism for Azobenzene in a Cholesteric Liquid Crystal Solvent," *J. Am. Chem. Soc.*, 100:5953.
7. Rau, H. and E. Luddecke. 1982. "On the Rotation-Inversion Controversy of Photoisomerization of Azobenzene: Experimental Proof of Inversion," *J. Am. Chem. Soc.*, 104:1616.
8. Winter, N. H. and R. S. Pitzer. 1975. "Theoretical Description of Diimide Molecule," *J. Chem. Phy.*, 62:1269.
9. Beveridge, D. L. and H. H. Jaffe. 1966. "The Electronic Structure and Spectra of cis- and trans-Azobenzene," *J. Am. Chem. Soc.*, 88:1948.
10. Patai, S. ed., 1975. *The Chemistry of Hydrazo, Azo and Azoxy Groups*, Part 1 & 2, New York, NY: John Wiley & Sons.
11. Lang, J. J., J. M. Robertson and I. Woodward. 1939. "X-Ray Crystal Analysis of Trans-Azobenzene," *Proc. Roy. Soc. London*, Sec. A 171:398.
12. Smets, G. 1983. "Photochromic Phenomenon in the Solid Phase," *Advances in Polymer Science*, Berlin: Springer Verlag, 50:17.
13. Williams, J. L. R. and R. C. Daly. 1977. "Photochemical Probes in Polymers," *Prog. Polym. Sci.*, 5:61–93.
14. Flory, P. J. 1939. "Kinetics of Polyesterification Effects of Molecular Weight and Viscosity on Reaction Rate," *J. Am. Chem. Soc.*, 61:3334.
15. Morawetz, H. 1978. "Comparative Studies of the Reactivity of Polymers and Their Low-Molecular-Weight Analogs," *J. Polym. Sci., Polymer Symposia*, 62:271.
16. Kuhn, W. 1934. *Kolloid-2*, 68:2.
17. Morawetz, H. 1977. *Contemporary Topics in Polymer Science*, E. M. Pierce and J. R. Schafgen, eds., New York, NY: Plenum, 2:171.
18. Tabak, D. and H. Morawetz. 1970. "Rates of Conformational Transitions in Solutions of Randomly Coiled Polymers. III," *Macromolecules*, 3:403.
19. Paik, C. S. and H. Morawetz. 1972. "Photochemical and Thermal Isomerization of Azoaromatic Residues in the Side Chains and in the Backbone of Polymers in Bulk," *Macromolecules*, 5:171.
20. Chen, D. T. L. and H. Morawetz. 1976. "Photoisomerization and Fluorescence of Chromophores Built in the Backbones of Flexible Polymer Chains," *Macromolecules*, 9:463.
21. Skolnick, J. and E. Helfand. 1980. "Kinetics of Conformational Transitions in Chain Molecules," *J. Chem. Phys.*, 72:5489.
22. Liao, T. P. and H. Morawetz. 1980. "Non-existence of Crankshaft-like Motions in Dilute Solutions of Flexible Chain Polymers," *Macromolecules*, 13:1228.
23. Kumar, G. S., P. Depra and D. C. Neckers. 1984. "Chelating Polymers Containing Photosensitive Functionalities," *Macromolecules*, 17:1912.

24. Kumar, G. S., P. Depra, K. Zhang and D. C. Neckers. 1984. "Chelating Polymers Containing Photosensitive Functionalities II," *Macromolecules*, 17:2463.
25. Lamarre, L. and C. S. P. Sung. 1983. "Studies of Physical Aging and Molecular Motion by Azochromophoric Labels Attached to the Main Chains of Amorphous Polymers," *Macromolecules*, 16:1729.
26. Irie, M. and W. Schnabel. 1985. "On the Dynamics of Photostimulated Conformational Changes of Polystyrene with Pendant Azobenzene Groups in Solutions," *Macromolecules*, 18:394.
27. Kamogawa, H. 1969. "Redox Behavior in Photochromic Polymers of Thiazine Series," *J. Appl. Polym. Sci.*, 13:1883.
28. Kamogawa, H., M. Kato and M. Sugiyama. 1968. "Synthesis and Properties of Photochromic Polymers of the Azobenzene and Thiazene Series," *J. Polym. Sci.*, 6:2967.
29. Kamogawa, H. 1971. "Synthesis and Properties of Photochromic Polymers of the Mercury Thiocarbazonate Series," *J. Polym. Sci.*, 9:335.
30. Kamogawa, H. and H. Hasegawa. 1973. "Redox Sensitization of the Photochromic Polymers of Thiazine Series," *J. Appl. Poly. Sci.*, 17:745.
31. Priest, W. J. and M. M. Sifain. 1971. "Photochemical and Thermal Isomerization in Polymer Matrices. Azo Compounds in Polystyrene," *J. Polym. Sci. Polym. Chem. Ed.*, 9:3161.
32. Sung, C. S. P., I. R. Gould and N. J. Turro. 1984. "Pulsed Laser Spectroscopic Study of the Photoisomerization of Azo Labels at Three Different Locations on a Polystyrene Chain," *Macromolecules*, 17:1447.
33. Yu, W. C., C. S. P. Sung and R. E. Robertson. 1988. "Site-Specific Labelling and the Distribution of Free Volume in Glassy Polystyrene," *Macromolecules*, 21:355.
34. Eisenbach, C. D. 1978. "Effect of Polymer Matrix on the Cis-Trans Isomerization of Azobenzene Residues in Bulk Polymers," *Makromol. Chem.*, 179:2489.
35. Eisenbach, C. D. 1979. "Photochromic Processes for the Investigation of Chain Mobility in Polymers," *Makromol. Chem.*, 180:565.
36. Eisenbach, C. D. 1980. "Cis-Trans Isomerization of Aromatic Azo Chromophores Incorporated in the Hard Segment of Poly (Ester-Urethanes)," *Makromol. Chem. Rapid Commun.*, 1:287.
37. Gronski, W., D. Emis, A. M. M. Bruderlin, M. M. Jacobi, R. Stadler and C. D. Eisenbach. 1985. "Deuterium Magnetic Studies on Strained Selectively Deuterated Elastomers," *Br. Polym. J.*, 17:103.
38. Williams, M. L., R. F. Landel and J. D. Ferry. 1955. "Temperature Dependence of Relaxational Mechanisms in Amorphous Polymers and Other Glass-forming Liquids," *J. Amer. Chem. Soc.*, 77:3701.
39. Kumar, G. S., C. Savariar, M. Saffran and D. C. Neckers. 1985. "Chelating Polymers Containing Photosensitive Functionalities—III," *Macromolecules*, 18:1525.
40. Stadler, R. and M. Webe. 1986. "Azo-Dye Junctions in Elastomeric Networks—I," *Polymer*, 27:1254.
41. Gronski, W., R. Stadler and M. M. Jacobi. 1984. "Evidence for Nonaffine and Inhomogenous Deformation of Network Chains in Strained Rubber-Elastic Networks by Deuterium Magnetic Resonance," *Macromolecules*, 17:741.

42. Sung, C. S. P., E. Pyun and H. L. Sun. 1986. "Characterization of Epoxy Cure by UV–Visible and Fluorescence Spectroscopy," *Macromolecules*, 19:2922.

43. Yu, M. C. and C. S. P. Sung. 1988. "Mobility and the Distribution of Free Volume in Epoxy Network by Photochromic Labelling and Probe Studies," *Macromolecules*, 21:365.

44. Irie, M. 1982. In *Molecular Models of Photoresponsiveness*, G. Montagnoli and B. F. Erlanger, eds., NATO ASI series, A68:291.

45. Bertelson, R. C. 1971. "Photochromism," *Techniques in Chemistry*, Glenn H. Brown, ed., New York, NY: Wiley Interscience, 3:795.

46. Irie, M., A. Menju and K. Hayashi. 1979. "Photoresponsive Polymers: Reversible Solution Viscosity Change of Poly(Methyl Methacrylate) Having Spirobenzopyran Side Groups," *Macromolecules*, 12:1176.

47. Irie, M., K. Hiramo, S. Hashimoto and K. Hayashi. 1981. "Photoresponsive Polymers 2: Reversible Solution Viscosity Change of Polyamides Having Azobenzene Residues in the Main Chain," *Macromolecules*, 14:262.

48. Blair, H. S., H. I. Pogue and E. Riordan. 1980. "Synthesis and Characterization of Inherently Colored Polyamides," *Polymer*, 21:1195.

49. Lovrien, R. and J. C. B. Waddington. 1964. "Photoresponsive Systems I: Photochromic Macromolecules," *J. Am. Chem. Soc.*, 86:2315.

50. Agolini, F. and F. P. Gay. 1970. "Synthesis and Properties of Azoaromatic Polymers," *Macromolecules*, 3:349.

51. Eisenbach, C. D. 1980. "Isomerization of Aromatic Azo Chromophores in Poly(ethyl acrylate) Networks and Photo-mechanical Effect," *Polymer*, 21:1175.

52. Irie, M. and H. Tanaka. 1983. "Photoresponsive Polymers. 5. Reversible Solubility Change of Polystyrene Having Azobenzene Pendant Groups," *Macromolecules*, 16:210.

53. Kungwatchakun, D. and M. Irie. 1988. "Photoresponsive Polymers. Photocontrol of the Phase Separation Temperature of Aqueous Solutions," *Makromol. Chem. Rapid Commn.*, 9:243.

54. Fowkes, F. M. 1964. *Contact Angle, Wettability and Adhesion*, ACS Adv. Chem. Series, No. 43, ACS, Washington DC.

55. Ishihara, K., N. Hawada, S. Kato and I. Shinohara. 1983. "Photoresponse of the Release Behavior of an Organic Compound by an Azoaromatic Polymer Device," *J. Polym. Sci. Chem. Ed.*, 21:1551.

56. Ishihara, K., N. Negishi and I. Shinohara. 1982. "Photocontrolled Adsorption Chromatography for Lysozyme Using Azoaromatic Polymer," *J. Appl. Polym. Sci.*, 27:1897.

57. Ishihara, K., N. Negishi and I. Shinohara. 1981. "Photoregulated Binding Ability of Azoaromatic Polymer for Surfactant," *J. Polym. Sci. Polym. Chem. Ed.*, 19:3039.

58. Irie, M. and R. Iga. 1987. "Photoresponsive Polymer Blends," *Makromol. Chem. Rapid Commun.*, 8:569.

59. Ishihara, K., S. Kato and I. Shinohara. 1982. "Separations of Proteins by Polymeric Adsorbents Containing Azobenzene Moiety as a Ligand," *J. Appl. Pol. Sci.*, 27:4273.

60. Erlanger, B. F. 1976. "Modes of Photoregulation," *Ann. Rev. Biochem.*, 45:267.

61. Goodman, M. and A. Kossoy. 1966. "Conformational Aspects of Polypeptide Structure. XIX," *J. Am. Chem. Soc.*, 88:5010.
62. Goodman, M. and A. Kossoy. 1967. "Conformational Aspects of Polypeptide Structure. XXIII," *J. Am. Chem. Soc.*, 89:3863.
63. Ueno, A., J. Anzai, T. Osa and Y. Kadoma. 1977. "Light-Induced Conformational Changes of Polypeptides," *J. Polym. Sci., Polym. Lett. Ed.*, 15:407.
64. Ueno. A., J. Anzai, T. Osa and Y. Kadoma. 1977. "Light-Induced Conformational Changes of Polypeptides," *Bull. Chem. Soc., Jpn.*, 50:2995.
65. Ueno, A., J. Anzai, T. Osa and Y. Kadoma. 1977. "Light-Induced Conformational Changes of Polypeptides: Photoisomerization of Azoaromatic Polypeptides," *Bull. Chem. Soc., Jpn.*, 52:549.
66. Ueno, A., J. Anzai and T. Osa. 1977. "Thermal Isomerization of Azoaromatic Residues in the Side Chain of Polypeptides," *J. Polym. Sci. Polym. Lett. Ed.*, 17:149.
67. Ciardelli, F., O. Pieroni and A. Fissi. 1986. "Photocontrol of the Solubility of Azobenzene-Containing Poly(L-Glutamic Acid)," *J. Chem. Soc. Chem. Commun.*, 264.
68. Ueno, A., K. Takahashi, J. Anzai and T. Osa. 1981. "Reversible Helix Reversion Induced by Light to Polyaspartates with Small Amounts of Photochromic Side Chains," *Chem. Lett.*, p. 113.
69. Ueno, A., K. Takahashi, J. Anzai and T. Osa. 1981. "Reversible Conformational Changes Induced by Light in Copolymers of Phenyl Azo Substituted Aspartic Acids," *Makromol. Chem.*, 182:693.
70. Ueno, A., Y. Morikawa, J. Anzai and T. Osa. 1984. "Conformational Changes in Solid Films among Helices with Different Helical Structures," *Chem. Lett.*, 1453.
71. Houben, J. L., A. Fissi, D. Baccila, N. Rosato, O. Pieroni and F. Ciardelli. 1983. "Azobenzene Containing Poly(L-glutamates) Photochromism and Conformation in Solution," *Int. J. Bio. Macromol.*, 5:94.
72. Atreyi, M., M. V. R. Rao and P. V.Scania. 1984. "Conformation of Poly(L-lysine) Containing Some Azoaromatic Side Chains," *J. Macromol. Sci. Chem.*, A21:15.
73. Fissi, A., O. Pieroni and F. Ciardelli. 1987. "Photoresponsive Polymers: Azobenzene-Containing Poly (L-Lysine)," *Biopolymers*, 26:1993.
74. Kinoshita, T., M. Sato, A. Tacizawa and Y. Trujita. 1986. "Photoinduced Reversible Conformational Transition of Polypeptide Solid Membranes," *Macromolecules*, 19:51.
75. Stoekenius, W., R. H. Lozier and R. A. Bogomolini. 1979. "Bacteriorhodopsin and the Purple Membrane of Halobacteria," *Biochim. Biophys. Acta*, 505:215.
76. Ermakova, V. D., V. D. Arsenov, M. I. Cherkashin and P. P. Kisilitsa. 1977. "Photochromic Polymers," *Russ. Chem. Rev.*, 46(2):145.
77. Shinkai, S., M. Ishihara and O. Hauabe. 1985. "Photoresponsive Crown Ethers. XVII." *Polymer J.*, 17(10)1141.
78. Shinkai, S., T. Nakaji, T. Ogawa, M. Shimomura and O. Mauabe. 1981. "Photoresponsive Crown Ethers of Photocontrol of Ion Extraction and Ion Transport by a Bis(crown ether) with a Butter-fly like Motion," *J. Amer. Chem. Soc.*, 103:111.

79. Shinkai, S., T. Ogawa, Y. Kusano, O. Hauabe, K. Kikukawa, T. Goto and T. Matsuda. 1982. "Photoresponsive Crown Ethers. 4," *J. Amer. Chem. Soc.*, 104:1960.
80. Shinkai, S., T. Nakaji, Y. Nishida, T. Ogawa and O. Hauabe. 1980. "Photoresponsive Crown Ethers I," *J. Am. Chem. Soc.*, 102:5860.
81. Anzai, J., H. Sasaki, A. Ueno and T. Osa. 1983. "Photo-Induced Potential Changes across Poly(vinyl chloride)-Crown Ether Membranes," *J. Chem. Soc., Chem. Commun.*, p. 1045.
82. Anzai, J., H. Sasaki, A. Ueno and T. Osa. 1984. "Poly(vinyl chloride)/Azobenzene Linked Bis(15-crown-5) Membranes," *Chem. Lett.*, p. 1205.
83. Anzai, J., A. Ueno, H. Sasaki, K. Shimokawa and T. Osa. 1983. "Photocontrolled Permeation of Alkali Cations through a Poly(vinyl chloride)/Crown Ether Membrane," *Makromol. Chem., Rapid Commun.*, 4:731.
84. Anzai, J., H. Sasaki, K. Shimokawa and T. Osa. 1984. *Nippon Kagaku Kaishi*, p. 338.
85. Kumano, A., O. Niwa, T. Kajiyama, M. Takayanagi, K. Kano and S. Shinkai. 1980. "Photoresponsive Polymers IX," *Chem. Lett.*, p. 421.
86. Kano, K., Y. Tanaka, T. Ogawa, M. Shimomura, Y. Okahata and T. Kunitake. 1980. "Photoresponsive Membranes. Regulation of Membrane Properties," *Chem. Lett.*, p. 421.
87. Okahata, Y., H. Lim and S. Hachiya. 1983. "Photoresponsive Capsule Membrane Permeability," *Makromol. Chem. Rapid Commun.*, 4:303.
88. Morgan, C. G., E. W. Thomas, Y. P. Yianni and S. S. Sandhu. 1985. "Incorporation of a Novel Photochromic Phospholipid Molecule into Vesicles of Dipalmitoyl Phosphatidylcholine," *Biochem. Biophy. Acta*, 820:107.

6

BIODEGRADATION OF AZO FUNCTIONAL POLYMERS

6.1. Introduction

At different times in the history of polymer science specific subjects have come to center stage for intense investigation, because they represent new and important intellectual challenges as well as technological opportunities. Two decades ago, the chief incentive for the scientific study of biodegradation of polymers was concern with the problem of preventing or retarding microbial attack on polymer surfaces, fibers, films, etc. Most polymers, by themselves, are resistant to microbial attack because of their hydrophobic character which inhibits the activity of extracellular enzymes. Often, the biodeterioration observed is due to the susceptibility of additives, such as plasticizers, to microbial fouling. Many research efforts in polymer chemistry were directed toward the synthesis of polymers resistant to biodegradation. In the late seventies, this early interest in polymer/environment interactions changed from a preoccupation with damage of plastics by biological agents to concern with pollution by discarded plastics, and since then there has been a raft of publicity about degradable plastics.

For many biomedical, agricultural and ecological purposes, it is preferable to have a biodegradable polymer that will undergo degradation by enzymes in the body or by bacteria in the soil. The term "biodegradable polymer" refers to a polymer which undergoes enzymatic degradation in a physiological or microbial environment and where the secretion of the degrading enzyme(s) is linked to a chemically recognizable cleavage of the molecular structure of the polymer.[1,2] Biodegradable polymers have found applications in medicine and surgery as absorbable sutures and drug delivery vehicles and in agriculture as mulches and seedling containers. The burden of accumulating waste has also given impetus to the cause of degradable plastics.

Biopolymers contain polar groups and decompose on heating before they melt and hence cannot be easily fabricated. On the other hand, most of the synthetic polymers, hitherto commercially exploited, are not biodegradable. To overcome this apparent paradox, several synthetic approaches have been formulated and tried with fair success. The most common strategy has been to incorporate hydrolyzable groups found in natural polymers such as aliphatic carboxylic and phosphoric esters, amide groups and the glycosidic linkages. Several groups of workers have reported novel polymers with the incorporation of various biodegradable "functional groups" (amides, esters, etc.), "units" such as aminoacids, and "blocks" such as peptide or oligosaccharide blocks in polymeric backbone, side chains and crosslinks which serve as "starters" to initiate enzymatic hydrolysis. Many of the functional groups investigated as susceptible linkages in these materials undergo degradation under aerobic conditions and very few polymeric structures have been specifically tailored for anaerobic conditions.

The *in vivo* azo reduction with the cleavage of aromatic linkage has received considerable attention because of its resistance to hydrolases and its sensitivity to anaerobic bacteria as established in the case of anti-inflammatory drugs such as prontosil and azulfidine.[3] Polymeric materials, whose degradation is specific to the anaerobic reductive microflora of the large intestine, are ideal candidates for orally delivering those drugs which are sensitive to digestive enzymes or which are rapidly absorbed. Likewise, anaerobic degradation of plastic waste in landfills is another area of continued interest. It is this special property of the azo functional group which has led to a new class of degradable materials in recent years.

6.2. Bacterial Azo Reduction

The *in vivo* azo reduction with the cleavage of the aromatic azo linkage is one of the earliest explored biological reactions of the functional groups.[3-10] The metabolic reduction is considered as an important detoxication route together with that of aromatic hydroxylation. The two stage azo reduction (**117**) to amines (**119**) proceeding by way of hydrazo (**118**) intermediates can be regarded as the reverse reaction of non-enzymic and enzymic formation from hydrazo compound and aromatic amines (see also Chapter 2).

The ability of the microflora to reduce the azo groups of various xenobiotic compounds has been known for many years.[3-10] However, the specific components of the intestinal microflora participating in azo reduction and the mechanism of these enzymatic processes are poorly

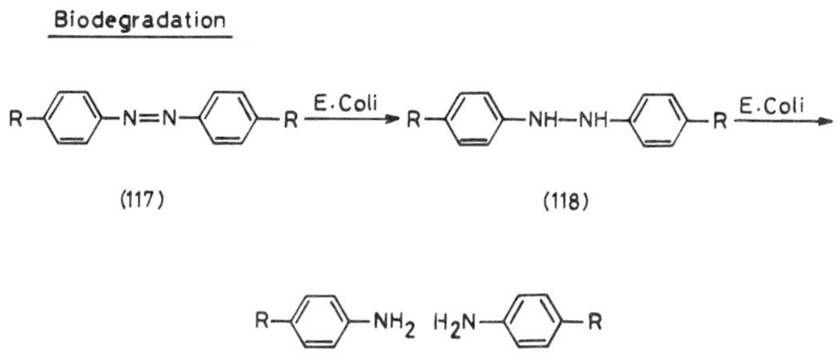

Biodegradation

$$R-C_6H_4-N{=}N-C_6H_4-R \xrightarrow{E.Coli} R-C_6H_4-NH{-}NH-C_6H_4-R \xrightarrow{E.Coli}$$

(117) (118)

$$R-C_6H_4-NH_2 \quad H_2N-C_6H_4-R$$

(119)

FIGURE 6.1

understood. The different rates of degradation of azo dyes are attributed mainly to the solubility of the dye rather than on any structural feature. Earliest reports were published by Sisley and Porcher[4] who obtained sulphanilic acid as a metabolite of Orange I (a phenyl azo naphthalene dye) after oral administration to dogs. Following the discovery in 1932 of the antibacterial activity of prontosil, the biotransformation to the active sulphanilamide was hypothesized by Trefouel et al.[5] and thereafter by Fuller.[6]

What seems to have sparked the interest among the polymer chemists to design azo-based polymer systems is the metabolic pathways of prontosil and azulfidine(120) as shown in Figure 6.2. Prontosil undergoes azo reduction to form sulphanilamide which is inhibitory against bacteria, and 1,2,4 triaminobenzene.[3] Likewise, azulfidine (SAS) was synthesized during World War II as a sulfonamide but found its widest use in the inflammatory diseases of the lower bowel. After a dose of SAS, very little of the unchanged drug is found in blood and urine (Figure 6.2), while the metabolites sulfapyridine(121) and 5-aminosalicylic acid(122) appear.[11,12]

Several workers[3] have clearly established that bacterial species that are common in the lower intestine of the rodent and mammalian animals are able to reduce the azo bond and split SAS.

6.3. Bacterial Reduction of Polymeric Azo Dyes

Intestinal bacteria under anaerobic conditions have also been reported to reduce high molecular weight polymeric derivatives of certain azo dyes at rates comparable with those of the low molecular weight parent com-

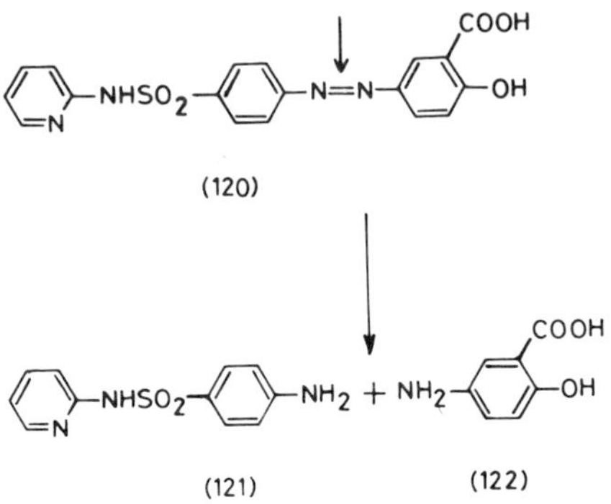

FIGURE 6.2 Metabolic reactions of azulfidine (SAS) in the lower gut.

pounds, despite large differences in the membrane permeabilities of these compounds. The polymeric azo dyes studies are based on the food dyes sunset yellow and tartrazine.[13-15] Parkinson et al.[13] have compared metabolism and intestinal absorption of polymeric derivatives of sunset yellow(123) and tartrazine(124) (Figure 6.3) in rats to metabolism and absorption of sunset yellow(125) and tartrazine(126) themselves. The sulfonamide group was chosen to link azo dye chromophore to the polyaminoethylene backbone of the polymeric dyes in part because of its apparent biological stability seen in antimicrobial sulfonamides.

The metabolic fate of orally administered {${}_{14}C$} polysulfonylamido ethylene(127), which is a common metabolite of the polymeric derivatives of both sunset yellow and tartrazine (Figure 6.3) suggested that the sulfonamide bond linking the azo dye chromophores to the polyaminoethylene backbone is extremely stable *in vivo*. Their work also concluded that azo bond cleavage is not inhibited significantly by attachment of the chromophore to the polymer backbone. Further studies of anaerobic reduction of mono- and polymeric dyes by intestinal bacteria *in vitro* have demonstrated that reduction involves extracellular electron mediators and is not dependent on penetration of the dye into the bacterial cell.[16] Azo reduction was observed by spectrophotometric measurement of visible absorption at λ_{max} value for each substrate dye. The ability to reduce polymeric dyes was not limited to a particular component of the intestinal microflora, but rather seemed to be shared by all bacteria tested, to a varying degree. Similar results are reported for low molecular weight azo dyes.[15] Also, the reduction of polymeric azo dyes by bacterial cell suspensions almost invariably was stimulated by a variety of low potential ($E = -200$ to -350 mV) electron carriers, particularly flavins such as FMN.[16] Again, similar observations have been made for low molecular weight azo dyes. These observations are the basis for the design of a new class of site-specific controlled release polymeric prodrugs for the therapy of lower bowel and other diseases such as ulcerative colitis.

The success of polymeric dyes as food candidates was short lived, attachment of chromophores to the high molecular weight polymer in no way guarantees biological inertness.[17] In this case, the azo bond reduction and accompanying chromophore cleavage were found to occur readily in the gastrointestinal tract (GI), presumably due to the presence of reductive cofactors of low molecular weight. Ironically, it is the problem with microbial reduction which led to the abandonment of polyazo dye as food additives and the search then turned to other chromophore classes which were not susceptible to reductive cleavage.

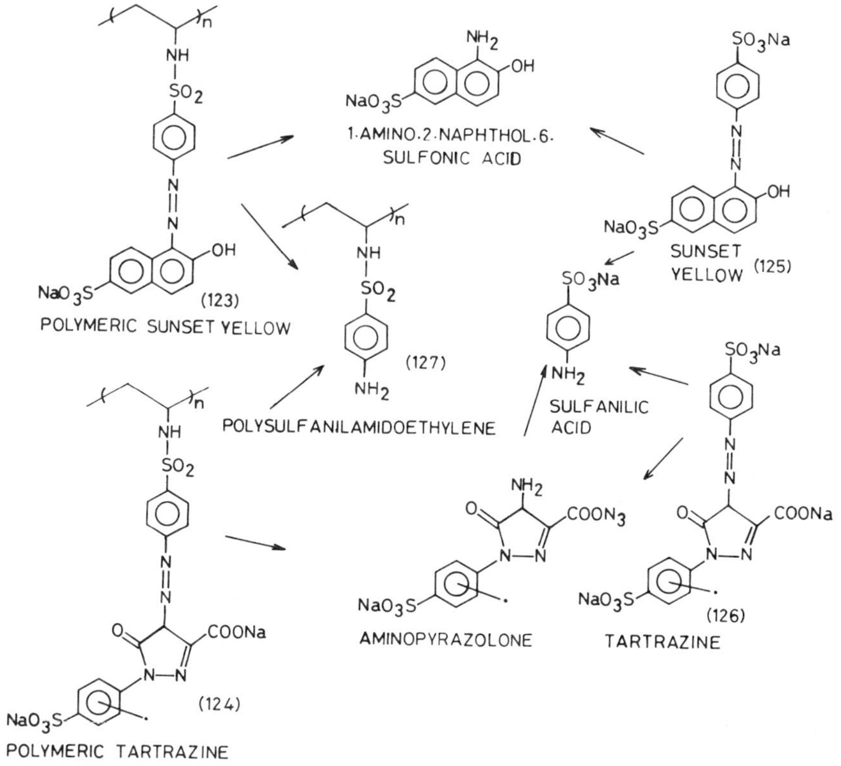

FIGURE 6.3 Metabolic reactions of polymeric derivatives of sunset yellow(123) and tartrazine(124) in mammalian lower gut.

6.4. Design of Biodegradable Aromatic Azo Linkage

For the optimization of azo polymers as oral drug delivery vehicles, it is important to establish *in vitro* bacterial degradation of azo polymers by the species present in the human intestine before any trails are made on the delivery of drugs. It is also important to correlate, with certain obvious limits, degradability with the structure of low molecular weight model substrates and polymers with the same degradable units in polymer backbone, side chains and crosslinks. It seems to us that our present knowledge of structure-degradability relationships of azo compounds is insufficient to permit design of biodegradable polymers on a rational basis. Based on the little information which is available so far, it is desirable to introduce the following structural features in designing degradable polymers.

(1) Azoaromatic units with electron-withdrawing substituents—such as SO_3Na groups to facilitate azo reduction[18]
(2) Substituted azobenzene and azonaphthalene units and a "bay region" which is known to enhance azo reduction[19]
(3) Azo moieties with heterocyclic azoaromatic units, for example, azo pyridine and azo imidazoles which reduce electron density on the nitrogen atoms[18]
(4) Spacer groups: the nature, polarity, and length of the spacer group separating the degradable bond from the backbone which has vital influence on bacterial susceptibility of the polymer[20,21]

6.5. Types of Biodegradable Drug Delivery Systems

The microbial susceptibility of polymeric dyes has led the scientists developing drug delivery systems to look for polymeric azo carriers for physical targeting of drugs. It must be pointed out that all the reports in this area have appeared only in the mid-eighties and information available is still incomplete. The drug delivery systems based on degradable carriers fall into three categories.[22-24] In the first category, the polymer acts as a chemical carrier where the pharmacologically active unit is chemically attached to the polymer thus acting as a prodrug and releases the amino-containing drug upon microbial reduction. The other approaches use azo polymers as physical carriers. From a chemical angle, a polymeric carrier can be synthesized following any of the following three mechanisms.

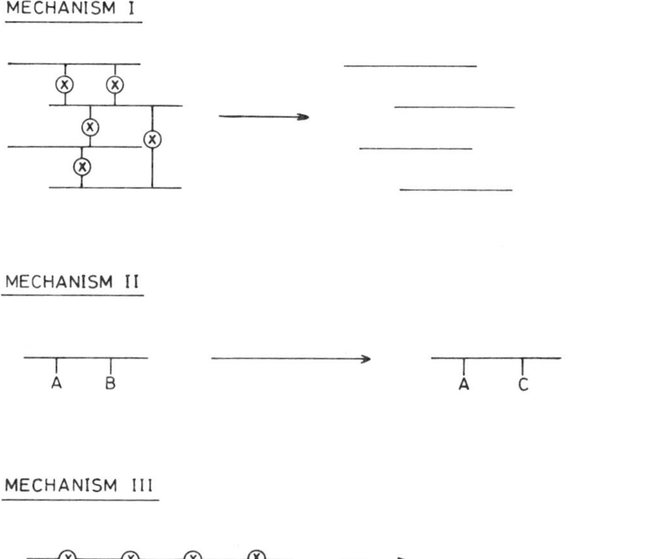

FIGURE 6.4 Schematic representation of polymer degradation mechanisms for drug delivery.

MECHANISM I

In these systems, water-soluble polymers or hydrophilic polymers are insolubilized by means of degradable crosslinks. Consequently, the polymer matrix is highly hydrophilic and completely permeated by water. Since the active agent is located in an aqueous environment, its water solubility becomes an important consideration and compounds with appreciable water solubility will be rapidly leached out independent of the matrix erosion rate. A useful application of this system is for the slow release of water soluble macromolecules such as peptide drugs that can be entangled and encapsulated in this system and are able to escape when a sufficient number of crosslinks have cleaved and the matrix crosslink density has been reduced.

MECHANISM II

Systems in this category include all polymers that are initially water-insoluble but become water-soluble as a consequence of hydrolysis, ionization or protonation of pendant groups. Because side chain cleavage results in little polarity and molecular weight changes, this mechanism is unlikely to serve as a useful system. These systems can serve as polymeric prodrugs when the liberated moiety due to enzymatic action is a pharmacologically active agent. In other words, these polymers can serve as chemical carriers for certain drugs.

MECHANISM III

This category includes hydrophilic polymers that undergo enzymatic or hydrolytic backbone cleavage and are solubilized by conversion to non-toxic low molecular weight compounds. Incorporation of the azoaromatic units has to be carefully planned because highly aromatic polymers tend to be insoluble. Besides, polymers should be designed in such a way so as not to leave any aromatic amines as degradation products.

6.6. Azo Polymers as Chemical Carriers: Degradable Side Chains

6.6.1. Site-Specific Delivery of Anti-Inflammatory Agents

When azulfidine (SAS) (Figure 6.2) is fed to conventional rats, none of the drug is recovered in the urine, feces or caecum; the excreata contain products of the azo bond reduction, sulfapyridine and 5-aminosalicylate and other metabolites in yields greater than 50%. Parkinson et al.[25,26] have reasoned that a pharmaceutical dosage form containing a polymer-leashed form of 5-ASA (Figure 6.5) would be more effective than SASP for the site-specific release of 5-ASA in the colon. The potential advantages of the polymeric drug include non-absorption in the small intestine and elimination of side effects caused by (121). The polymer has to be a non-metabolizable, non-toxic polymer to which is attached a precursor of 5-ASA. These acid or salt groups are attached to the backbone through an azo linkage between a backbone aromatic carbon and 5-position carbon on the salicylate. When the polymer passes through the mammalian GI tract, azo links are bacterially cleaved releasing 5-ASA. The generation of 5-ASA occurs in the lower bowel where its therapeutic action is desired. The backbone polymer itself passes through the gut substantially intact and unabsorbed.

Parkinson et al.[25,26] have considered numerous polymers which can serve as chemical carriers for the anti-inflammatory drugs with or without spacer groups separating the azo linkage from the main chain. Some of the representative structures are shown in Figure 6.6. Although the details of

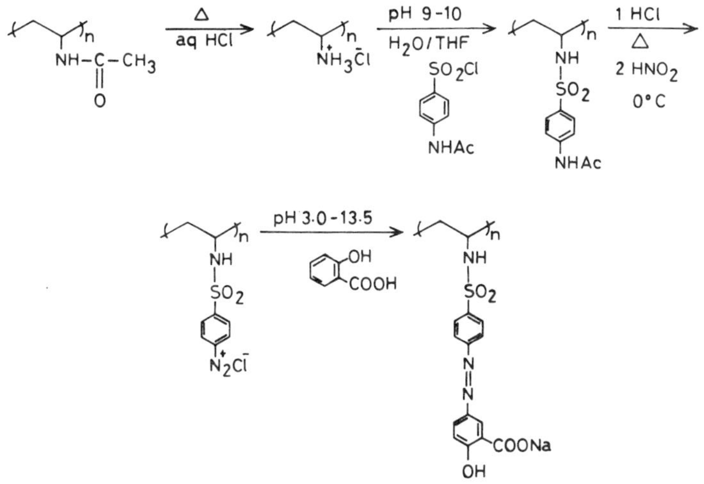

(128)

FIGURE 6.5 Preparation of polymer-bound salicylic acid as a polymeric prodrug.

in vitro and *in vivo* experiments are not detailed in the patent applications filed by Parkinson et al., it is to be anticipated that the backbone probably will have to maintain some amount of hydrophilicity so that the azoreductases can get a "foothold" on the polymer for the reductive cleavage. It is unlikely to be very efficient when the azo group is linked to a very hydrophobic backbone. Likewise, the spacer group separating the degradable bond is also very important. It is known that a length of 3-4 methylene units is ideal for the efficient cleavage to occur with other hydralyzable bonds.

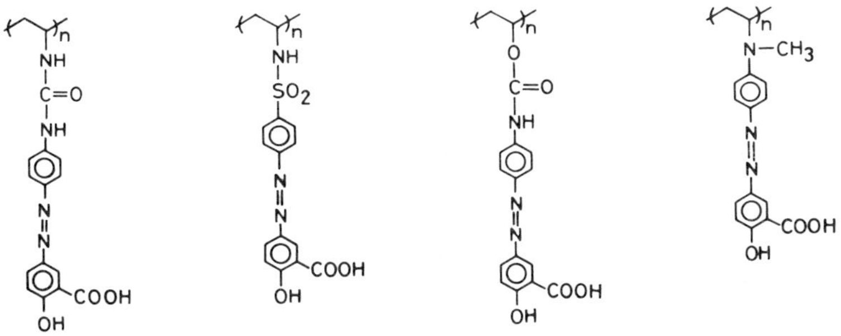

(129) (130) (131) (132)

FIGURE 6.6

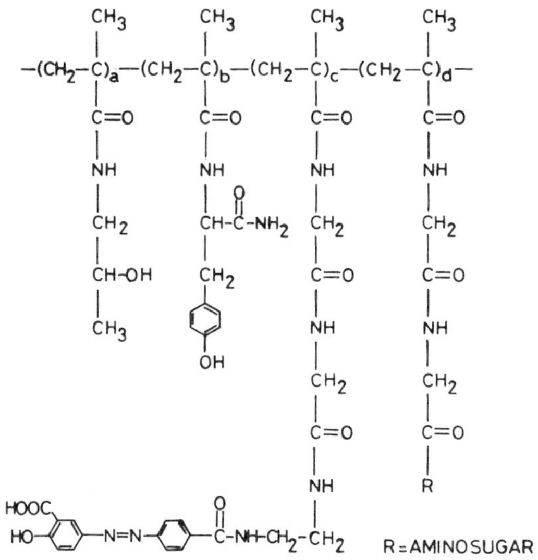

R = AMINOSUGAR

(133)

FIGURE 6.7

6.6.2. *Bioadhesive Polymeric Carriers for Colon Delivery*

One of the major problems associated with the site-specific oral drug delivery is the short gastrointestinal residence time of most available drug formulations. Water-soluble polymers have been proposed for colon-specific delivery of 5-ASA and loosely crosslinked charged polymers have been proposed to control GI transit time. Recently, bioadhesive water-soluble HPMA copolymers which combine the concept of site-specific drug delivery and bioadhesion have been synthesized by Kopecek and coworkers[27,28] and release of 5-ASA from these copolymers by the action of microbial azoreductases was demonstrated *in vitro*.

Their approach involves the copolymerization of hydroxypropyl methacrylate (HPMA) with polymerizable derivatives of drug 5-ASA and bioadhesive moieties. This route provides a better control over purity, molecular weight distribution and loading of drug as well as bioadhesive moieties (up to 80 mole%) to the polymers. Bioadhesive HPMA copolymers used in their earlier studies were synthesized by radical copolymerization of HPMA, MA-Tyr-NH_2, and MA-Gly-Gly-ONp followed by incorporation of bioadhesive carbohydrate moieties and drug by consecutive aminolysis. However, only up to 15 mole% of side chains containing reactive p-nitrophenyl ester groups could be incorporated by copolymerization due to the fact that the reactive comonomer (MA-Gly-Gly-ONp) participates in the chain transfer reaction during copolymerization. Consequently, the amount of drug and bioadhesive carbohydrate moieties which could be incorporated into the copolymer was limited. The alternative route involving copolymerization of functionalized monomers of both azo and sugar moieties permits the synthesis of HPMA copolymers(133) with high contents of bioadhesive carbohydrate moieties. Using this method, HPMA copolymers with different content of fucosylamine were synthesized. To determine the site of binding, bioadhesive properties of HPMA copolymer to the GI tract of female adult guinea pigs were evaluated *in vitro* using everted sacs of small intestine and colon with and without inhibitors, etc. (Figure 6.8).

6.7. Azo Polymers as Physical Carriers

The advantages to millions of diabetics of oral administration are so great that much effort was directed just after the discovery of insulin by Banting and Best[29] to the development of alternatives to injection but without success. Two apparent barriers to oral administration of insulin are (1) the inactivation of insulin, a 51-aminoacid polypeptide, by digestion in the gastrointestinal tract, and (2) lack of knowledge of a mechanism to transport peptides with more than three aminoacid residues from the gut to the blood. While synthetic and recombinant DNA technologies make available an almost limitless variety of peptides, the delivery systems of these drugs have not grown with equal vigor. For this application, it is possible in principle to design polymers with degradable bonds either in the backbone or in the crosslinks of polymers as shown in Figure 6.4. In the last few years, both these approaches have been reported.

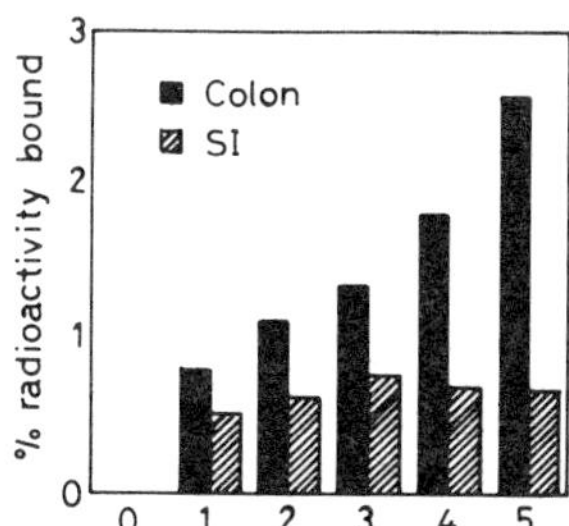

FIGURE 6.8 Binding of ^{125}I-labelled hydroxypropyl methacrylate (HPMA) copolymers to everted sacs from guinea pig small intestine and colon. 0—control, 1—2.2% Fucosylamine (FucN); 2—3.2% FucN; 3—9.9% FucN; 4—18.1% FucN; 5—33.2% FucN (30 minutes incubated at 37°C).

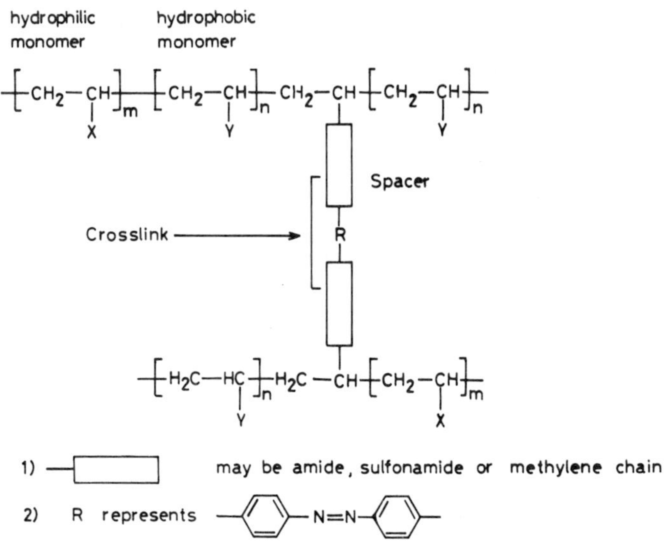

FIGURE 6.9 A schematic representation of azo crosslinked polymers employed for oral delivery of peptides.

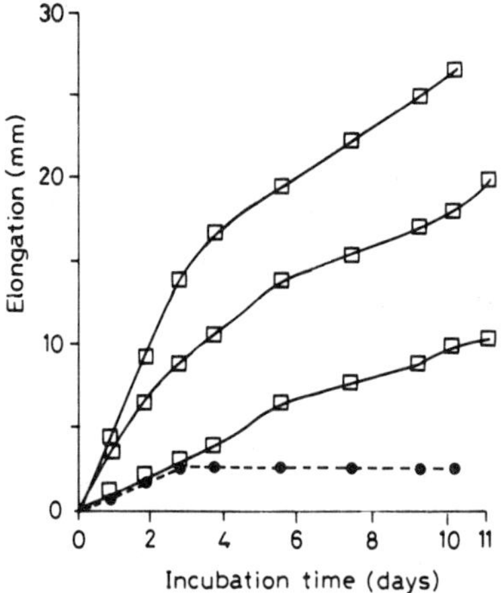

FIGURE 6.10 Increase in length of 10 mm threads of azoaromatic crosslinked STY–HEMA copolymer during anaerobic incubation in medium with (□) and without inoculation (●) of human feces. The upper curves were from three other threads of the same polymer, incubated in inoculated medium. Because of variation in the thickness of the thread (0.07 to 0.25 mm) and density of the inoculum, the results are not combined.

6.7.1. Azo Crosslinked Hydrogels: Oral Insulin?

Saffran, Kumar, Neckers and coworkers[30-33] have approached the problem of peptide delivery using azo crosslinked hydrogel carriers. As a new approach to oral delivery, peptide drugs were coated with polymers crosslinked with azoaromatic groups (Figure 6.9) to form an impervious film to protect orally administered drugs from ingestion in the stomach and small intestine. The crosslinkers were designed with amide, sulfonamide spacer groups. When the azo polymer coated drug reached the large intestine, the indigenous microflora reduced the azo bonds, broke the crosslinks, and degraded the polymer film, thereby releasing the drug into the lumen of the colon for local action or absorption. In *in vitro* testing, azoaromatic polymer threads were incubated in anaerobic cultures of human fecal microflora and the elongation with a 2 g load was measured over 11 days of incubation. Threads in uninoculated medium elongated by a mean of only 12.5% in the first 3 days as the polymer absorbed some water, but threads exposed to bacteria continued to elongate at a faster rate, reaching two to almost four times their original length (Figure 6.10).

The ability of the azo polymer coating to protect and deliver orally administered peptide drugs was demonstrated in rats with the peptide hormones vasopressin and insulin. Figure 6.11 illustrates a typical experiment

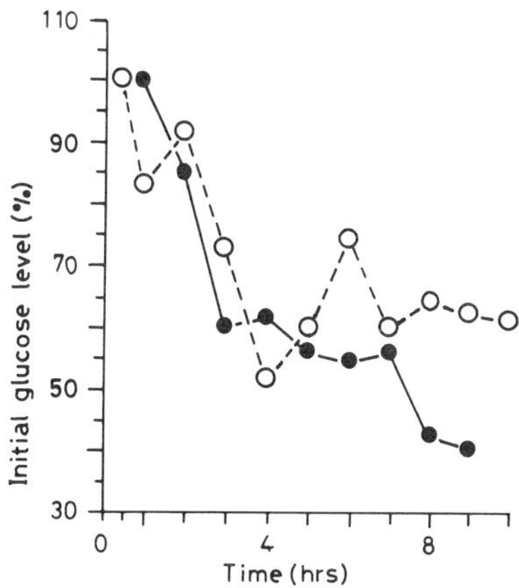

FIGURE 6.11 The effect on blood glucose of the oral administration of an azoaromatic polymer-coated pellet containing 1 IU (about 28 mmol kg body weight) of insulin on the blood glucose levels in two rats made diabetic with strepotozotocin. The initial blood glucose levels were 400 mg per 100 ml (○ --- ○) and 290 mg per 100 ml (● —— ●) respectively. The falls in blood glucose became significant (decrease > 15%) 3 hrs after the administration of insulin.

in which oral insulin caused a sustained fall in the blood glucose in two diabetic rats. There was no decrease in blood glucose over the same period in diabetic rats given control pellets. The action of orally administered insulin in normal rats was less marked.

These results show that the system operates in principle. However, the responses occurred at variable times, from 1 to 9 hrs after the administration of the protected peptide; therefore, results with different animals cannot be combined. There is no explanation at present for the variation in time of delivery of the peptides. In some animals that did not respond for 10 hours or more, the pellets were found in the stomach debris upon autopsy. Thus hold up in the stomach may explain some of the variation. From direct observation of transit time of a dye-solution in water-loaded rats, the minimal transit time of liquids to the ileo-cecal junction was about 1 hr. However, solids, such as polystyrene microspheres, have been known to reach the end of the small intestine with a transit time of 195 minutes.

Many characteristics of this new delivery system remain to be established. These include the relation between the chemical composition and the physical properties of the azoaromatic polymers and their ability to deliver drugs to the colon, the effect of the nature of azocoating on the pellets and the passage of the drugs from the site of release in the intestine to the blood. Until a combination of an animal model and an azoaromatic polymer with a reproducible delivery time is found, bioavailability studies will be difficult to carry out.

6.7.2. *Large-Intestine Degradable Polyurethanes*

Following the approach of azo crosslinked polymers, Kimura et al.[34] have reported several segmented polyurethanes (Figure 6.12) having azoaromatic groups in the main chain, which are degraded by the action

Azo Segment (from below) : $(O-R^1-OCONH-R^2-NHCO)_x$

PEG Segment (hydrophilic) : $[O(CH_2CH_2O)_n CONH-R^2-NHCO]_y$

PG Segment (hydrophobic) : $[OCH_2CH(CH_3)OCONH-R^2-NHCO]_z$

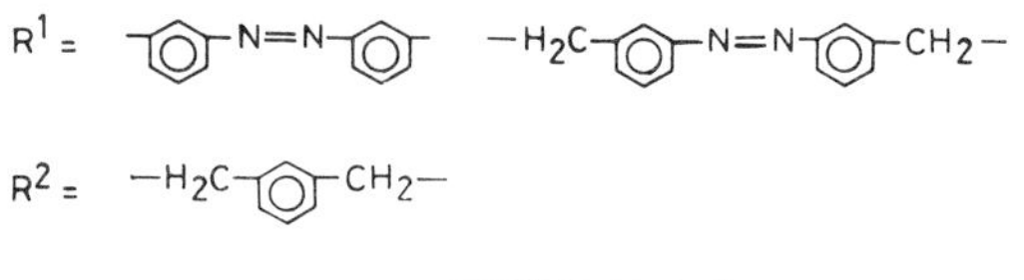

R^1 = $-C_6H_4-N=N-C_6H_4-$ $\quad -H_2C-C_6H_4-N=N-C_6H_4-CH_2-$

R^2 = $-H_2C-C_6H_4-CH_2-$

FIGURE 6.12

of intestinal flora. The polyurethane films and the drug pellets coated with the polyurethanes were prepared and incubated anaerobically in the general anaerobic media with and without human intestinal flora to investigate the polymer degradation and the drug release.

After incubation of the films, most of the azo groups of the polyurethanes were found to be reduced to hydrazo groups without any change in the molecular weight. The strength and elongation of the films also was much lower and degradation was much higher with decreasing molecular weight of the polyurethanes.

The rate of drug release from the pellets was much faster in the flora media than in flora-free media. The pellets coated with the polymers containing 10 mole% of the azoaromatic groups exhibited a higher degradation rate than those coated with the ones containing 5 mole% of azo content. It was therefore concluded that the rate of the drug release could be controlled by changing both the PEG ratios in the polyurethanes. The authors also reported an *in vivo* test using beagle dogs and it was observed that the drug release started when the coating pellets reached the ascending colon and had been almost completed before they reached the descending colon.

6.8. Polymers of Controlled Lifetime

Development of novel polymers of controlled lifetime, whose outdoor and indoor stability can be tuned at molecular level, has been a long cherished goal for packaging applications.[35]

Several approaches towards photo- and biodegradation of polymers are reported. A novel approach by Mahalingam et al.[36] is based on the combination of photo and biodegradation incorporating azoaromatic units susceptible to anaerobic microflora and photosensitive ketone groups. This methodology offers a unique in-built photodegradative mechanism which can be triggered by anaerobic azoreductases of the bacteria. The key to the design is the selection of azobenzene moiety whose function is two-fold. Azobenzene and its derivatives undergo a reversible photochemical isomerization. Since all ketone polymers absorb around 280–290 nm, the azo chromophore may provide UV shielding to ketone groups.[34] Secondly, many bacteria possess the ability to reduce the azo compounds and the reduction has wide species and substrate specificity. Hence, the incorporation of azo functional monomers in STY-MVK copolymers should increase the outdoor stability of these ketone polymers. Upon disposal, these pale yellow polymers are expected to be cleaved by the azoreductases of soil bacteria opening the “absorption window” to ketone groups.

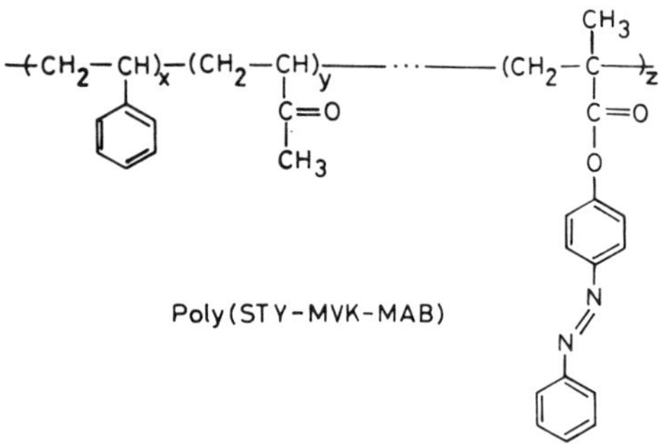

Poly(STY-MVK-MAB)

(134)

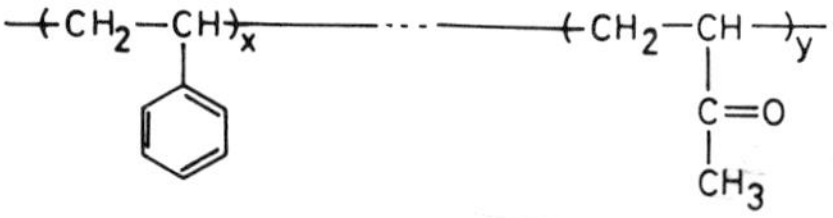

Poly(STY-MVK)

(135)

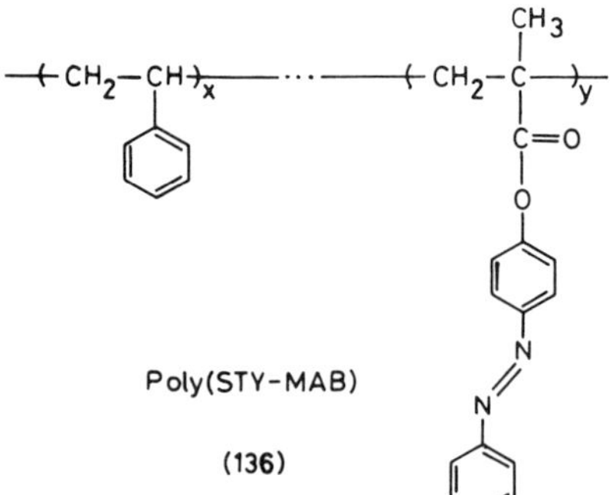

Poly(STY-MAB)

(136)

FIGURE 6.13

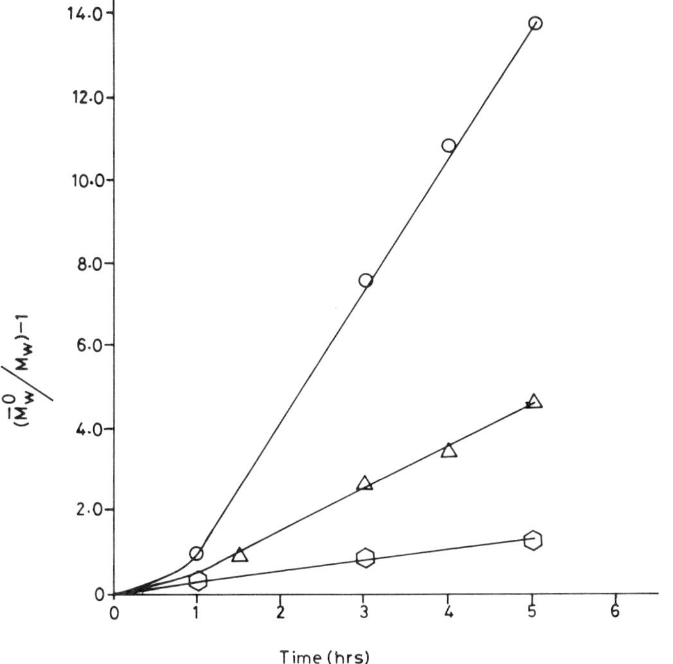

FIGURE 6.14 UV shielding effect of azoaromatic sidechains. Photolysis of polymers in toluene (△) STY-MVK-MAB(134); (○) STY-MAB(135); and STY-MVK (– ⬡ –). Temp – 30°C.

The polymers investigated (134,135,136) are shown in Figure 6.13. The azo moiety exhibits a strong absorption maximum at 325 nm. The polymers (134) and (136) undergo *trans-cis* photoisomerization upon irradiation. The intensity of absorption at 352 nm progressively decreases with irradiation time and λ_{max} shifts towards shorter wavelengths. Depending on the photostationary composition, the azo moieties may mask the ketone groups reducing the amount of energy absorbed by the ketone groups. The UV shielding effect of MAB was measured as an index for the chain cleavage and the results are presented in Figure 6.14. The molecular weight of the polymers was measured as an index for the chain cleavage and the results are presented. In comparison to STY-MVK copolymer, STY-MAB-MVK terpolymer (1% MAB) is significantly more resistant to UV degradation. Control experiments with STY-MAB copolymer indicated little or no change in molecular weight.

Microbiological experiments with *E. coli* indicated about 20–30% reduction in the intensity of the azo absorption after a week of incubation under anaerobic conditions. For the same period, aerobic degradation was much less. Soil burial tests of these polymers and the photodegradation after the azo cleavage are important in tailoring the lifetimes of these polymers which have not yet been reported.

6.9. Conclusion

In this chapter, I have tried to summarize the known facts relating to the biodegradation of azo functional polymers. The special property of anaerobic bacterial degradation of azo linkages has been utilized in polymer science only in the last few years. It seems that present knowledge in this area is insufficient to permit the full exploitation of the challenges it provides. The knowledge of the mechanism of biodegradation and the structural parameters which influence it in the case of low molecular weight analogs would allow one to synthesize polymers which are not only suitable as drug carriers, but also suitable for applications in which the desired lifetime could be adjusted according to the degradation profile. Much of the information available is for water-soluble azo dyes which have been in traditional use for long. All these dyes are based on azobenzene and azonaphthalene ring systems. No information is available on azo heterocycles as well as aliphatic azo compounds. The beginning of activity in this area, still rather minor, is encouraging and eventually may reach its potential value.

6.10. References

1. Kumar, G. S., V. Kalpagam and U. S. Nandi. 1982/83. "Biodegradable Polymers—Prospects and Progress," *J. Macromol. Sci. Rev. in Macromol. Chem. & Phys.*, 22(2):225.
2. Kumar, G. S. 1986. *Biodegradable Polymers—Prospects and Progress*, New York, NY: Marcel Dekker Inc., p. 83.
3. Miyadera, T. 1974. *The Chemistry of Hydrazo, Azo and Azoxy Groups*, Part I, S. Patai, ed., New York, NY: Wiley & Sons, p. 502.
4. Sisley, P. and C. Porcher. 1911. *C.r.hebd. Seanc. Acad. Sci., Paris*, 152:1062.
5. Trefouel, J., M. J. Trefouel, F. Nitti and D. Bovet. 1935. *Soc. Biol.*, 120:756.
6. Fuller, A. T. 1937. "Is p-Aminobenzenesulfonamide the Active Agent in Prontosil Therapy?" *Lancet.*, 1:194.
7. Walker, J. 1970. "Metabolism of Azo Compounds; A Review of the Literature," *Food. Cosmet. Toxicol.*, 8:659.

8. Daniel, J. W. 1962. "The Excretion and Metabolism of Edible Food Colors," *Toxicol. Appl. Pharmacol*, 4:572.
9. Scheline, R. R. and B. Longberg. 1965. "The Absorption, Metabolism and Excretion of the Sulfonated Azo Dye, Acid Yellow, by Rats," *Acta Pharmacol. Toxicol.*, 23:1.
10. Ryan, A. J. and P. G. Welling. 1967. "Enzymatic Reduction of Tartrazine by Proteus Vulgaris from Rats," *Food. Cosmet. Toxicol.*, 5:645.
11. Schroeder, H. and D. E. S. Campbell. 1972. "Absorption Metabolism and Excretion of Salicylazosulfapyridine in Man," *Clin. Pharmacol. Therap.*, 13:539.
12. Peppercorn, M. A. and P. J. Goldman. 1972. "The Role of Intestinal Bacteria in the Metabolism of Salicylazosulfapyridine," *Pharmacol. Exper. Therap.*, 181:555.
13. Honohan, T., F. E. Enderlin, B. A. Ryerson and T. M. Parkinson. 1977. "Intestinal Absorption of Polymeric Derivatives of the Food Dyes Sunset Yellow and Tartrazine in Rats," *Xenobiotica.*, 7:765.
14. Brown, J. P. 1981. "Reduction of Polymeric Azo and Nitro Dyes by Intestinal Bacteria," *Appl. Environ. Microbiol.*, 41:1283.
15. Dubin, P. and K. L. Wright. 1975. "Reduction of Azo Food Dyes in Cultures of Proteus Vulgaris," *Xenobiotica.*, 5:563.
16. Chung, K. T., G. E. Fulk and M. Egan. 1976. "Azo Dye Reduction by Anaerobes," *Appl. Environ. Microbiol.*, 35:558.
17. Dawson, D. J. 1981. "Polymeric Dyes," *Aldrichimica. Acta*, 14:2.
18. Walker, R. and A. J. Ryan. 1971. "Some Molecular Parameters Influencing Rate of Reduction of Azo Compounds by Intestinal Microflora," *Xenobiotica.*, 1:483.
19. Fujita, S., M. Suzuki and T. Suzuki. 1984. "Structure-Activity Relationships in the Induction of Hepatic Drug Metabolism by Azo Compounds," *Xenobiotica.*, 14:565.
20. Kopecek, J. and K. Ulbrich. 1983. "Biodegradation of Biomedical Polymers," *Prog. Polym. Sci.*, 9:1.
21. Goldberg, E. P. 1978. In *Polymeric Drugs*, L. Donaruma and O. Vogl, eds., New York, NY: Academic Press, p. 239.
22. Heller, J. 1980. In *Controlled Release of Bioactive Materials*, New York, NY: Academic Press, p. 1.
23. Lee, P. I. 1988. *Controlled Release Systems—Fabrication Technology*, CRC Press, p. 61.
24. Ratner, B. D. and A. S. Hoffman. 1976. In *Hydrogels for Medical and Related Applications*, J. D. Andrade, ed., ACS Symposium Series, Vol 31, Washington DC, ACS.
25. Parkinson, T. M., J. P. Brown and R. E. Wingard. U.S. Pat. 4,190,176, Feb. 20, 1980.
26. Parkinson, T. M., J. P. Brown and R. E. Wingard. U.S. Pat. 4,298,595, Nov. 3, 1981.
27. Rathi, R. R., B. Rihova, P. Kopeckova and J. Kopecek. 1991. "Bioadhesive Water Soluble Polymeric Drug Carriers for Site-Specific Drug Delivery," *Polymer Science: Contemporary Themes,*" S. Sivaram, ed., New Delhi: Tata-McGraw Hill, p. 1032.

28. Fornusek, L., Y. A. Grim, R. Duncan, J. F. Woodley and J. Kopecek. 1989. *Proc. Intl. Symp. Controll. Bioact. Materials*, 16:398.

29. Bliss, M. 1982. *The Discovery of Insulin*, Chicago: University of Chicago.

30. Kumar, G. S., C. Savariar, M. Saffran and D. C. Neckers. 1985. "Chelating Polymers Containing Photosensitive Functionalities," *Macromolecules*, 18:1525.

31. Saffran, M., G. S. Kumar, C. Savariar, J. C. Burnham, F. Williams and D. C. Neckers. 1986. "A Novel Approach to the Oral Administration of Insulin and Other Peptide Drugs," *Science*, 233:1081.

32. Saffran, M., G. S. Kumar, D. C. Neckers, R. H. Jones and J. B. Field. 1990. "Biodegradable Azopolymer Coating for Oral Delivery of Peptide Drugs," *Biochem. Soc. Trans.*, p. 752.

33. Saffran, M., C. Bedra, G. S. Kumar and D. C. Neckers. 1988. "Vasopressin: A Model for the Study of Effects of Additives on the Oral and Rectal Administration of Peptide Drugs," *J. Pharma. Sci.*, 77:33.

34. Kimura, Y., T. Kumagai, H. Yamane and T. Kitao. 1990. "A Novel Drug Delivery System by Using Large Intestine-Degradable Polyurethanes," *Proc. of Intl. Symposium on Polymer Supported Reactions*, Kyoto, Japan, Sept. 20–24, p. 83.

35. Guillet, J. E. 1973. *Polymers and Ecological Problems*, New York, NY: Plenum Press, p. 1.

36. Mahalingam, H. and G. S. Kumar. 1989. "A Novel Photo- and Biodegradable Plastic," *Abstracts of ACS Meeting*, Dallas, Texas, BTEC42.

7

MISCELLANEOUS SPECIALITY PROPERTIES OF AZO FUNCTIONAL POLYMERS

7.1. Introduction

Significant progress in synthesizing and processing organic polymers has led to an ever growing materials base. These developments along with theoretical advances, have led to an increasing knowledge of structure-property relationships. The newest step in this ladder of increasing sophistication is the creative integration of this knowledge into the designing of systems wherein tailoring of the micro- and macrostructure of polymers is extended to advanced final structures.[1] The earlier chapters have dealt essentially with the chemical properties of the azo functionality. In this chapter are outlined some of the speciality applications of azo polymers, which are based on the physical characteristics such as molecular geometry, color, etc. of the azo functional unit. Polymers in industry are on the verge of expanding beyond their long established passive roles to dynamic new applications. Many of these reports are very recent and represent probably what is to come in the coming decade.

7.2. Polymeric Azo Dyes

The origins of a polymeric dye are no doubt lost in antiquity along with the name of the first chemist to synthesize one. Macromolecular dyes are a group of compounds thought of as intrinsically or structurally colored polymers, i.e., possessing inherent properties. Their applications are markedly dependent on their relatively high molecular weight. Polymer-bound dye systems are of interest in photochemical research, in photographic processes and in a number of biologically related applications. For many of these applications, polymeric dyes should be water-soluble to give a sparkling clear solution, should have a high tinctoral strength, and be ionic or charged. The subject of macromolecular dyes has been reviewed by Dawson,[2] Guthrie,[3] Asquith[4] and by Marechal.[5] Broadly speaking, three different strategies have been followed to prepare polymeric azo dyes.

7.2.1. Polymerization of Colored Azo Monomers

This approach has been reported by several workers to be a viable one. Often this route takes the form of derivatizing a commercially available dye with a polymerizable group; an example is making acrylate esters or methacrylamide derivatives such as (137,138,139).

For example, Hrabak et al.[6] have prepared a series of new polymerizable azo dyes containing the methacryloyloxy groups. Their melting points, yield, color, stability and UV spectra were determined and the relationship between the properties and the structure was elucidated. The characteristics of homopolymerization of one of the azo dyes, N-methyl-N-{4-[2,5,dichlorophenyl]azophenyl}-2-aminoethyl methacrylate and of its copolymerization with a number of vinyl monomers were determined. By using the seeded polymerization of the azo dye in the industrially produced butadiene-styrene (35/65) latex, a colored latex was produced.

7.2.2. Polycondensation

Polycondensation is another route popular in the literature. In its usual form, a dye containing two amino-groups(140) is condensed with a bis acidchloride(141) or a diisocyanate to give a colored condensation polymer(142).

7.2.3. Modification of Preformed Polymers

Polymer backbones may be electrophilic or nucleophilic in character. Nucleophilic polymers provide the easiest route to macromolecular dyes largely because of the scarcity of double electrophiles which can induce desirable crosslinking. Conversely, there is an abundance of nucleophiles, e.g., H_2O, capable of crosslinking a polyelectrophile. The general principles of attaching the dye to the polymer are represented in Figure 7.3.

The physical and chemical properties of these macromolecular dyes can be altered by the inclusion of comonomers such as acrylonitrile, acrylic acid, and acrylate monomers which change the hydrophilic/hydrophobic nature of the resulting polymer. A number of acrylated azo dyes have also been grafted onto fibers such as cellulose, polypropylene, etc.[7] Many of these applications may be considered as conventional applications which

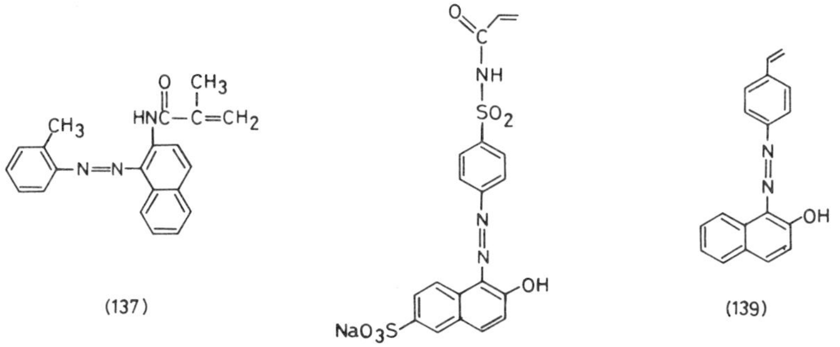

FIGURE 7.1

are based on physical characteristics and fall beyond the scope of this chapter, but the following two examples indicate the "special effects" that the presence of azo linkage imparts to polymeric dyes.

Polyphosphazenes(143) are appealing carrier molecules for chromophores because of their general transparency, photochemical stability of the backbone, the possibility of the electronic interactions between the backbone and the chromophore, and the ease with which different side chains can be incorporated onto −P−N− backbone by substitution techniques. Allcock and coworkers[8] have prepared polymer-bound azo dyes by diazotization of high polymeric poly(organophosphazene)skeleton, followed by coupling to phenol, naphthal, and (p-aminophenyl)naphthalene. These reactions were preceeded by model compound studies. The aminophenoxy side groups were generated by reduction of 4-nitrophenoxy groups with PtO_2 and hydrogen as shown in Figure 7.4. The backbone was unaffected by the reduction, diazotization, and diazo coupling process. The generation of intense colors following the diazo coupling step provided a firm indication for the formation of structures (144) shown in Figure 7.4.

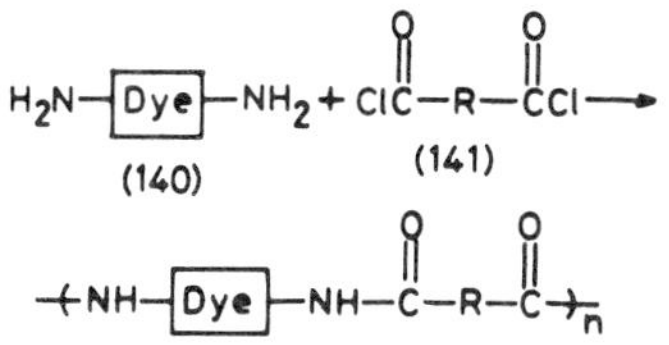

FIGURE 7.2

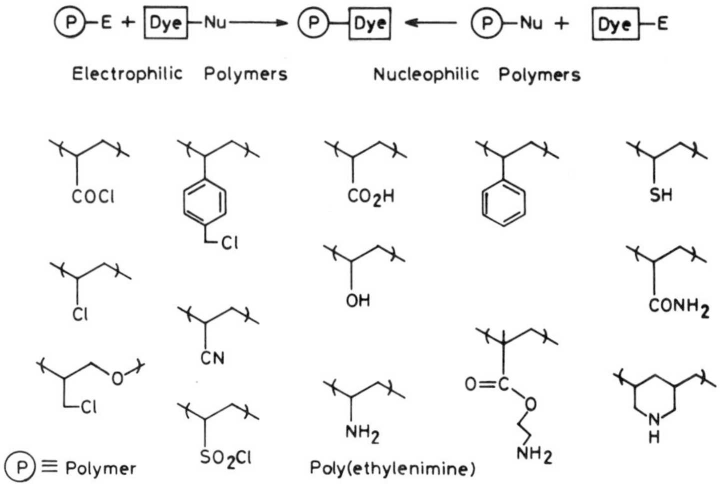

Electrophilic Polymers

Nucleophilic Polymers

COCl

Cl

CN

SO2Cl

CO_2H

OH

NH_2

Poly(ethylenimine)

SH

$CONH_2$

NH_2

N H

P ≡ Polymer

FIGURE 7.3

250 °C

(143)

$NaOC_6H_5$

-NaCl

$NaOC_6H_4NO_2$

-NaCl

H_2

(PtO_2)

HONO

HCl

HR

-HCl

(144)

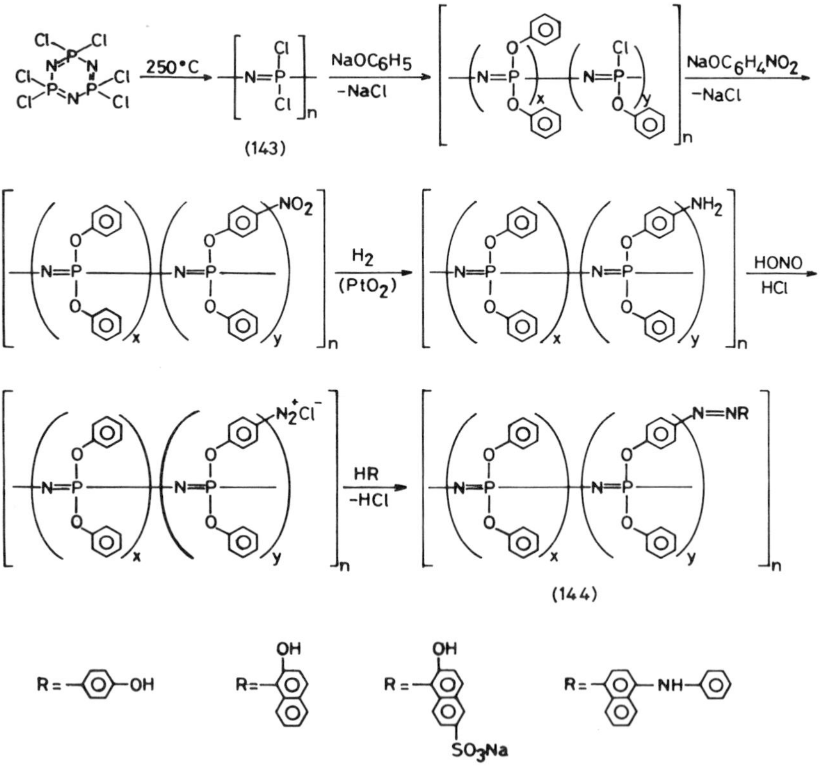

FIGURE 7.4 Preparation of poly(phosphazenes) containing azoaromatic side chains.

Another interesting special effect coloration incorporating azo groups is reported by Patel.[9] The principle of addition of color is well known. One can obtain a number of different color shades, for example, yellow/yellowish green, green/blue, upon gradual addition of a blue solution into a yellow solution.

Synthesis and polymerization of a large number of diacetylenes(145) with different functional substituent groups have been reported in the literature. For some applications of diacetylenes where more than one color is needed, it is possible to achieve it by synthesizing diacetylenes with chromophoric substituents. Unusual multiple color transitions can arise upon the polymerization of diacetylenes having colored substituents.

Diacetylenes with azobenzene side chains(145) polymerize in solid state either upon thermal annealing or upon exposure to high energy radiation as shown in Figure 7.6. The resultant polymers have fully conjugated backbones. Poly(diacetylenes)(146) are highly colored because the electrons are delocalized over long distances. Typically, the monomers are colorless solids while the partially polymerized diacetylenes are either blue or red. As the polymerization proceeds, the blue or red color intensifies. At high polymer yields (10%), the intense blue or red color starts appearing metallic gold or metallic green in reflectance.

Development of systems which incorporate macromolecular dyes has been somewhat uncoordinated and based largely on the desired applications. Application areas include textile dyes, paper colorants, food dyes, colorants for cosmetics, fugitive strains for registration purposes, immunoassay procedures, fluorescent brighteners, etc.

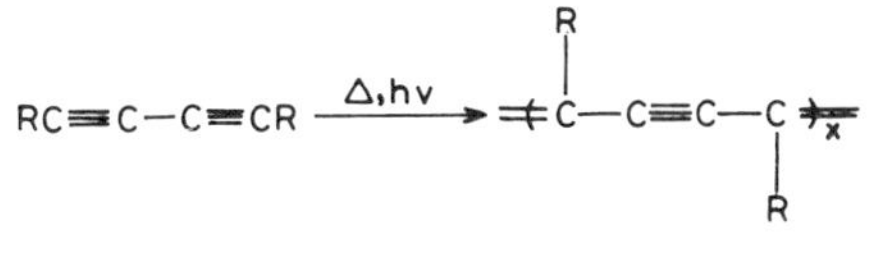

$$RC{\equiv}C{-}C{\equiv}CR \xrightarrow{\Delta, h\nu} {=}\!\!\left[\overset{R}{\overset{|}{C}}{-}C{\equiv}C{-}\underset{R}{\underset{|}{C}} \right]\!\!{=}_x$$

(145) (146)

$R{=}(CH_2)_n{-}O{-}CONHC_6H_4{-}N{=}N{-}C_6H_5$

or

$(CH_2)_n{-}O{-}SO_2{\cdot}C_6H_4{\cdot}N{=}N{-}C_6H_5$

Where $n = 1 - 4$

FIGURE 7.5 Polymerization of diacetylenes.

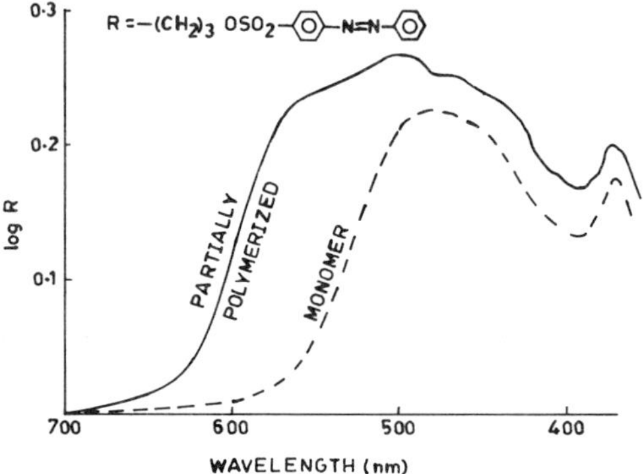

FIGURE 7.6 Reflectance spectra of sulfonate substituted diacetylene: (- - -) monomer; (——) partially polymerized by UV light. The spectra of the thermally polymerized sample is identical with that polymerized by UV light. The spectrum of the partially polymerized sample is expanded slightly for clarity.

7.3. Liquid Crystalline Polymers: Made to Order

The study of liquid crystal (LC) properties in polymer systems holds out much promise for the future, not only because such systems will provide us with new knowledge about both polymers and liquid crystals, but also from the stand point of their technological importance.[10] LC polymers come in several types. (1) The entire chain is mesogenic. (2) Mesogenic groups are in the side chains. (3) The mesogenic groups are in the main chain. Although the p-phenylene azo link is reported to favor LC behavior in small molecules, it has only recently been reported to cause LC polymer formation.[10]

7.3.1. Mesogenic Backbones

Polyesters have been the most popular type of thermotropic LC polymers. Hall et al.[11] have synthesized a number of aromatic and aliphatic polyesters from bisphenols containing one to three p-phenylene azo groups. These polymers have been examined for their LC properties. Aromatic polyformals based on multiazobisphenols have also been reported by the same authors. The bisphenol monomers used in this study are shown in Figure 7.7. These monomers were examined for their LC behavior. They were also converted to diacetates, diisobutyrates, mostly new compounds which were also examined for their LC behavior as models of the polymers.

An important factor in these studies was the tractability of the resulting polymers. Hall et al.[11] have used relatively soft diacyl halides to avoid the problem of intractability of bisphenol-based polymers. Because the soft diacid components present a trade off between LC character and tractability. Isophthaloyl chloride was the aromatic chloride employed. All the polyesters were synthesized by interfacial polycondensation. Polyformals were synthesized by the use of excess dichloromethane in N-methyl pyrollidone solution. Data was obtained on polymer melting behavior, inherent viscosities, film-forming ability, color and LC behavior. Polymer solubility was increased by the 5-t-butyl group in isophthalic acid and the methoxy in (**149**). Films were cast from polymer solutions. The occurrence of LC behavior depended strongly on the bisphenol monomer and on the diacid. Both small molecules and polysebacates based on (**150**) and (**153**) uniformly showed LC characteristic melting point behavior. Polysebacates based on (**148,149**) and (**151**) also showed LC properties. On the other hand, monomer or derivatives of (**151**) showed on LC behavior. All these LC compounds showed nematic textures.

The results agree with the notion that extended rod-like character is required for LC behavior. Only the sebacate group, with its ability to adopt the *trans* planar extended zig-zag form meets this criterion. The isophthalate residue, as well as the methylene formal units, lead to kinks in the chain. These disrupt the propensity of the LC character shown by the monomeric bisphenols and their simple derivatives. The aryl azo link,

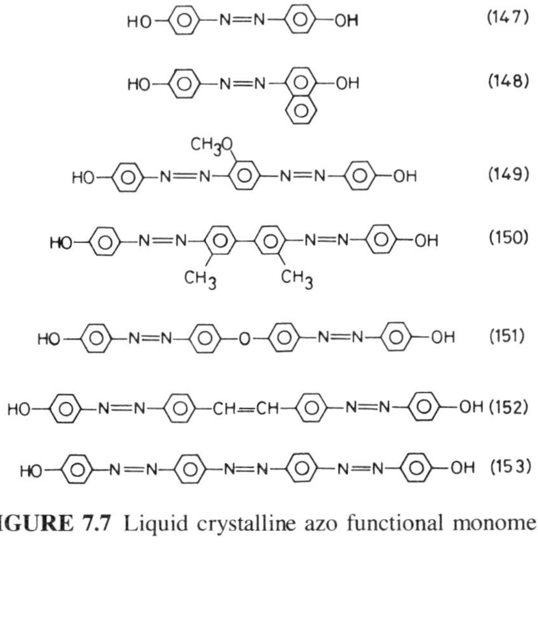

HO–⟨O⟩–N=N–⟨O⟩–OH (147)

HO–⟨O⟩–N=N–⟨O⟩–OH (148)

HO–⟨O⟩–N=N–⟨O⟩(OCH₃)–N=N–⟨O⟩–OH (149)

HO–⟨O⟩–N=N–⟨O⟩(CH₃)–⟨O⟩(CH₃)–N=N–⟨O⟩–OH (150)

HO–⟨O⟩–N=N–⟨O⟩–O–⟨O⟩–N=N–⟨O⟩–OH (151)

HO–⟨O⟩–N=N–⟨O⟩–CH=CH–⟨O⟩–N=N–⟨O⟩–OH (152)

HO–⟨O⟩–N=N–⟨O⟩–N=N–⟨O⟩–N=N–⟨O⟩–OH (153)

FIGURE 7.7 Liquid crystalline azo functional monomers.

in combination with sebacate units has been shown to form thermotropic liquid crystalline polyesters.

7.3.2. Mesogenic Side Chains

For liquid crystalline side chain polymers all relevant types of mesophases such as nematic, cholestric, and smectic have been realized. These polymers combine well-known properties of low molar mass liquid crystals with those of polymers and are of interest. A large number of polymers such as polymethacrylates, polyacrylates and polysiloxanes containing mesogenic pendant groups have been reported.[12-15] The enhancement of liquid crystalline properties by the polymerization of low-molecular weight liquid crystals was reported by Finkelmann et al.[15] Thermotropic liquid crystalline polymers were prepared by introducing flexible spacers between the polymer main chain and the rigid mesogenic side chain which decouple the motions of the mesogenic groups from the polymer main chain. Polyacrylate homo- and copolymers containing pendant substituted phenyl azo groups show liquid crystalline nematic phases. New enantiotropic liquid crystalline compounds based on 4-(p-substituted phenyl azo) phenol and m-cresol esters containing an acryloxy group were synthesized by Naikwadi et al.[16] Polyacrylates were prepared from these monomers by radical polymerization. The resulting polymers exhibit nematic enantiotropic properties. The polyacrylates have broader nematic ranges and higher nematic to isotropic transition temperatures compared to corresponding monomers. The nematic range is dependent on both the lateral substituents and terminal substituents. The laterally substituted polyacrylates gave a narrower nematic range than unsubstituted compounds. Increase in carbon number in the terminal alkyl group led to a broader nematic range. The LC polyacrylates could be coated uniformly on the inside of a glass capillary tubing. The coated capillary columns showed excellent selectivity in GC for the separation of isomeric compound. The LC stationary phases were stable up to 230°C without column bleeding.

The electro-optical effect, a color intensity change, in a guest-host display is based on the cooperative alignment of the guest dye in the nematic host and the different absorption of the dye depending on its orientation. Mixtures of dyes in liquid crystalline side chain polymers have been investigated and show similar orientation and electric field behavior as low molar mass systems. Colored polymeric liquid crystals with properties comparable to low molecular weight guest-host systems can be obtained by the copolymerization of mesogenic monomers and the monomeric dyes. They might open interesting application possibilities in the field of guest-

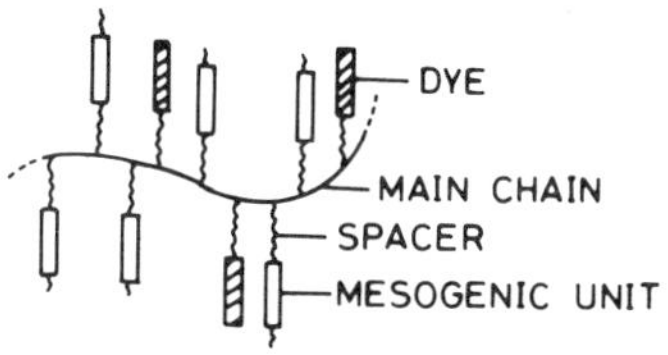

FIGURE 7.8 Schematic structure of dye containing liquid crystalline side group copolymers.

host systems. The schematic structure of such a dye containing liquid crystalline copolymers is shown in Figure 7.8.

By covalent fixation of both the dye and the mesogenic group to the same polymer chain it is possible to synthesize systems with high temperature independent dye concentrations. This is one of the main distinctions between the low molecular weight guest-host systems in which the solubility of the dye in the nematic host is frequently insufficient and temperature dependent. Ringsdorf et al.[17] have reported colored liquid crystalline polymers illustrated in Figure 7.9 (154,155). Copolymers with an azo dye content up to 40% weight were synthesized. The liquid crystalline behavior of the resulting polymers was studied by polarization microscopy and DSC measurements. All polymers showed a nematic mesophase as

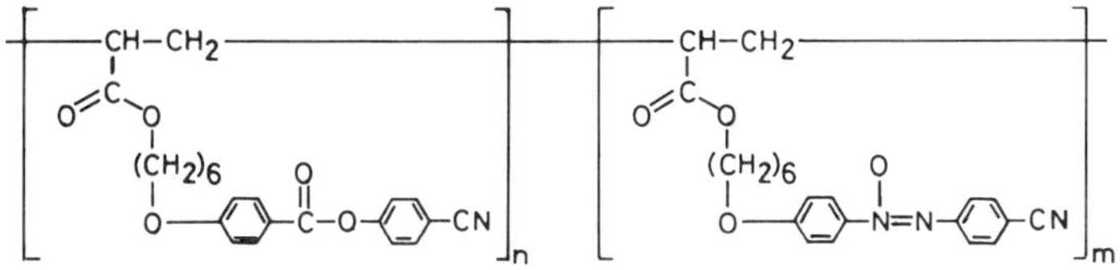

(154)

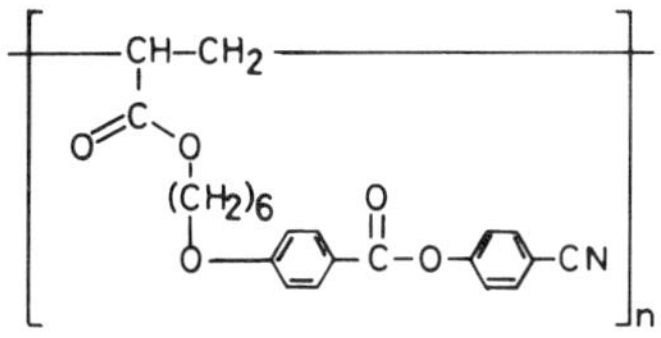

(155)

FIGURE 7.9

detectable by characteristic droplets at the nematic-isotropic transition. The DSC curves also showed that the phase behavior of the polymers is not influenced by their composition. In addition, the electro-optical effects of these copolymers and their mixtures of low molecular weight liquid crystals were investigated. These copolymers showed the same electro-optical effects as low molecular weight guest-host systems. Their macroscopic oriented nematic structure could be frozen in below the T_g resulting in a polymer film with dichroic properties. The behavior of the copolymers (phase behavior, surface and electric field orientation) could be improved via mixtures with low molar mass liquid crystals.

Use of polymer liquid crystals (PLC) for image storage materials has been reported by several groups.[18-20] Photon-mode image storage in the PLC was first demonstrated by Eich and Wendorff as "holographic" optical storage.[21,22] In their system, photoirradiation caused isomerization of the photochromic groups (azobenzene derivatives) incorporated into the PLCs, inducing simultaneous "grating" in the PLC. The "grating" was constructed essentially by the change in the refractive index resulting from the photoisomerization. In order to explore photochemically induced isothermal phase transition behaviors of PLCs with mesogenic phenyl benzoate and azobenzene side chains in relation to their morphological properties, orientational ordering of the PLCs has been studied recently by Ikeda et al.,[23,24] by means of colorimetric measurements and order parameter (S) determinations by FT-IR dichromism. Ikeda's system is different from the earlier approaches in that photochemically induced isothermal phase transitions of the matrix PLCs is the principle involved in the image storage system. These studies are important to construct highly efficient image storage systems based on PLCs.

7.4. Polymers with Non-Linear Optical Effect

Polymers that possess high non-linear optical activities have seen a recent growth in interest. Hall et al.[25] have recently synthesized polyesters containing multiple p-phenylene azo groups, which have strongly delocalized electrons. The introduction of donor and acceptor substituents in a push-pull arrangement to these polymers through the diphenyl azo group as shown in Figure 7.10 is expected to make them highly attractive as polymeric non-linear optical substrates.

Similarly, these dipolar units have also been incorporated into the side chains of comb-like polymethacrylates. Methacrylate or acrylate polymers containing dipolar azo groups in the side chain have been reported to exhibit a large third-order or second-order non-linear optical effect. In these

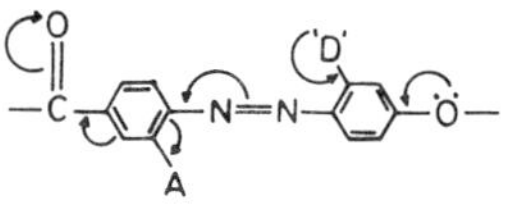

FIGURE 7.10

cases, the azo linkage acts more like a conduit than an acceptor. This is evident from the IR spectra which shows the shift of carbonyl group to lower frequency at 1680 cm^{-1}, indicating that the electrons are delocalized into the carboxylic acid group.

Monomer(156) was polymerized by a direct polycondensation in the presence of triphenylphosphine to give a red-brown polymer(157). DSC showed the polymer to have T_g at 141 °C followed by decomposition at 322 °C. The monomer(159) was prepared in a manner similar to the main-chain monomers. A six carbon spacer was used between the rigid polarizable group and the methacrylate unit. The monomer(159) was polymerized by free radical polymerization using AIBN as the initiator.

For non-linear optics, the wavelengths of importance are the two laser diode frequencies 1.3 μm and 830 nm. Many telecommunication applications use these frequencies. The other wavelengths of importance are 1.91, 1.34 and 1.06 μm. These polyesters(157) and polymethacrylates(160) might be of interest for non-linear optical applications.

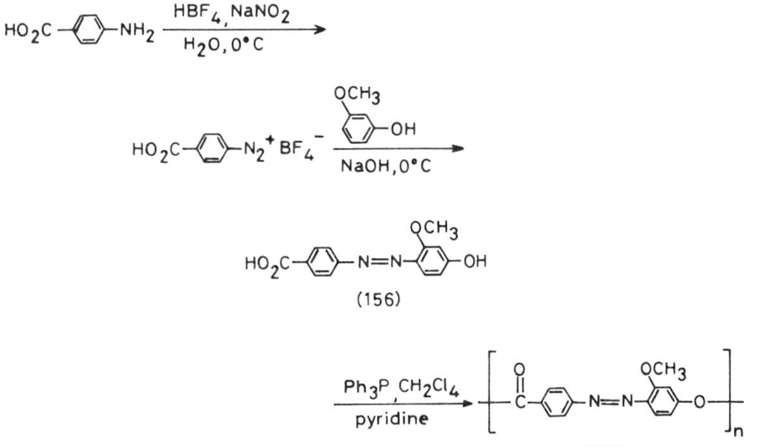

FIGURE 7.11 Synthesis of 4-[(4-hydroxy-2-methoxyphenyl)azo] benzoic acid.

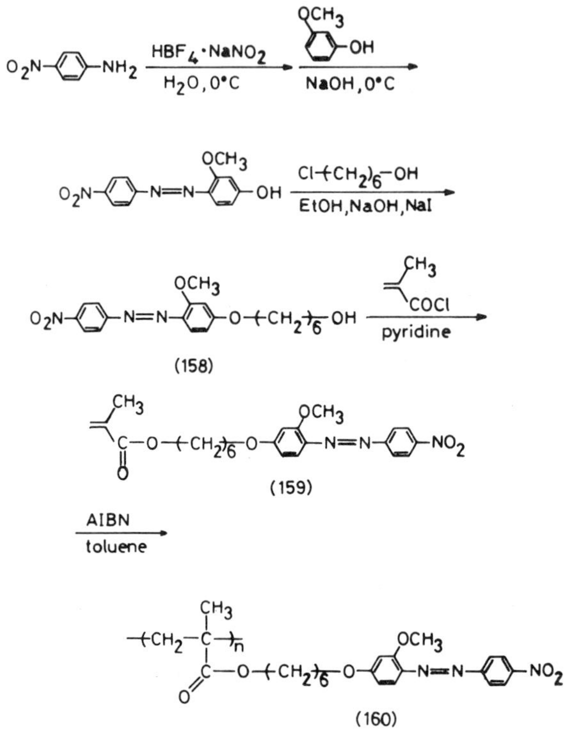

FIGURE 7.12 Synthesis of comb-like polymethacrylate containing dipolar p-phenyleneazo units in the side chains.

7.5. Conducting Polymers

Conjugated polymers based on stilbene derivatives, such as polyphenylene vinylene and aromatic polyimines such as polyaniline have been shown to be highly conducting upon doping. Although polymers containing p-phenylene azo links were reported long ago, their electrical properties have begun to be examined only now. The conductivity of such polymers was mostly investigated in the form of compressed pellets, owing to the intractibility of the polymers. Hall et al.[26] prepared films of azoaryl polymers and copolymers and investigated their electrical conductivities.

Various azopolymers were synthesized by the oxidative coupling of the aromatic diamines (Chapter 3). The presence of azoxy groups was evidenced in the IR spectra which showed peaks at 1230 cm^{-1} and elemental analysis. Selected polyazoarylenes were cast into films both from the polymerization solution and from the solutions obtained by dissolving the isolated polymer powders in trifluoroacetic or methanesulfonic acids. The solvent evaporated at room temperature or slightly elevated temperature. Thin films were purified by extracting with acetonitrile and ethanol for several days. The films were black with a metallic luster. Several films of copolymers were flexible and strong, and could be drawn (5%) and oriented on a hot bar. The polymers were n-doped by sodium naphthalide in THF or p-doped by iodine vapor. After the films became black with a metallic luster, the weight increased. Upon doping, the films became semiconducting generally showing conductivities of 10^{-4} to 10^{-5} S/cm in several days for n-doped film and in several hours for p-doped films. The n-doped films were more stable than the p-doped films perhaps due to the fact that reduction is easier than oxidation for these polymers.

7.6. Conclusion

So far, polymers have found widespread use in the electronics industry as insulating materials. These end-uses represent passive applications in the sense the polymer does not play an active role in the performance of the device. But active applications, in which the device performance depends on the active participation of the polymer, are increasing. Active applications for azopolymers include liquid crystalline polymers and polymers for non-linear optical effects. These areas constitute the research frontiers of polymer science. While it is unlikely that a new polymer will be introduced with the cost advantage and impact of polyethylene or nylon, advances will be made in speciality applications of polymers, where the unique properties of the polymers are central to the overall success of new technologies.

7.7. References

1. Tanner, D., V. Gabara and R. Schaefgen. 1988. In *Internl. Symposium on Polymers for Advanced Technologies*, M. Lewin, ed., Mainz, VCH, p. 384.
2. Dawson, D. J. 1981. "Polymeric Azo Dyes," *Aldrichimica Acta*, 14:32.
3. Guthrie, J. T. 1986. "Macromolecular Dyes," in *Encyclopedia of Polymer Science & Engineering, 2nd Ed.*, New York, NY: Wiley & Sons, 5:277.
4. Asquith, R. S., H. S. Blair, A. A. Crangle and E. Riordan. 1977. "Self-Colored Polymers Based on Anthraquinone Residues," *J. Soc. Dyers. Color.*, 93:114.
5. Marechal, E. "Polymeric Dyes: Synthesis, Properties and Uses," *Prog. Org. Coat.*, 10:251.

6. Hrabak, F., L. Chuiko, G. Voloshin, E. Morozova, E. Eliseeva and J. Lokaj. 1987. "Polymeric Azo Dyes," *Acta Polymer*, 38:643.
7. Selli, E. and I. R. Bellobono. 1981. "Photochemical Grafting of Acrylated Azo Dyes onto Polymeric Surfaces–III," *J. Soc. Dyes Color*, 97:438.
8. Allcock, H. R., P. E. Austin and T. F. Rakowsky. 1981. "Diazo Coupling Reactions with Poly (Organophosphazenes)," *Macromolecules*, 14:1622.
9. Patel, G. N. 1981. "Synthesis and Polymerization of Diacetylenes Having Chromophoric Substituent Groups," *Macromolecules*, 14:1170.
10. Ciferri, A., W. R. Krigbaum and R. B. Meyer. 1982. *Polymer Liquid Crystals*, New York, NY: John Wiley & Sons, p. 1.
11. Hall, H. K., Jr., T. Kuo, R. W. Lenz and T. M. Leslie. 1987. "New Polyesters and Polyformals Containing Multiple p-Aryleneazo Groups: Liquid Crystal Polyazoaryl Sebacates," *Macromolecules*, 20:2041.
12. Finkelmann, H., H. Ringsdorf and J. H. Wendorff. 1978. "Polyreactions in Ordered Systems–XIV," *Makromol. Chem.*, 179:273.
13. Ringsdorf, H. and R. Zentel. 1982. "Liquid Crystalline Side Chain Polymers and Their Behavior in the Electric Field," *Makromol. Chem.*, 183:1245.
14. Finkelmann, H., H. Ringsdorf, W. Siol and J. H. Wenderhoff. 1978. "Enantiotropic (Liquid Crystalline) Polymers: Synthesis and Models," *ACS Symposium Ser.*, 74:22.
15. Ringsdorf, H. and A. Schneller. 1982. "Liquid Crystalline Side Chain Polymers with Low Glass Transition Temperatures," *Makromol. Chem. Rapid Commun.*, 3:557.
16. Naikwadi, K. P., A. L. Jadhav, S. Rokushika and H. Hatano. 1986. "Synthesis of Liquid Crystalline Polyacrylates and Their Use in Capillary Gas Chromatography," *Makromol. Chem.*, 187:1407.
17. Ringsdorf, H. and H. W. Schmidt. 1984. "Electro-Optical Effects of Azo Dye Containing Liquid Crystalline Copolymers," *Makromol. Chem.*, 185:1327.
18. Coles, H. J. and R. Simon. 1985. "Electro-Optical Effects in Liquid Crystalline Side Chain Polymers," *Polymer*, 26:1801.
19. Sasaki, A. 1986. "Laser Beam Addressed Liquid Crystal Display," *Mol. Cryst. Liq. Crys.*, 139:103.
20. Pinsl, J., C. Brauchle and F. H. Kreuzer. 1987. "Liquid Crystalline Polysiloxanes for Optical Write-Once Storage," *J. Mol. Electro.*, 3:9.
21. Eich, M. and J. H. Wendorff. 1987. "Erasable Holograms in Polymeric Liquid Crystals," *Makromol. Chem. Rapid Commun.*, 8:467.
22. Eich, M. and J. H. Wendorff. 1987. "Reversible Optical Data Storage Using Polymeric Liquid Crystals," *Makromol. Chem.*, 8:59.
23. Ikeda, T., S. Horiuchi, D. B. Karanjit, S. Kurihara and S. Tazuke. 1990. "Photochemically Induced Thermal Phase Transition in Polymer Liquid Crystals with Mesogenic Phenyl Benzoate Side Chains–I," *Macromolecules*, 23:36.
24. Ikeda, T., S. Horiuchi, D. B. Karanjit, S. Kurihara and S. Tazuke. 1990. "Photochemically Induced Isothermal Phase Transition in Polymer Liquid Crystals with Mesogenic Phenyl Benzoate Side Chains–II," *Macromolecules*, 23:42.

25. Hall, H. K., Jr., T. Kuo and T. M. Leslie. 1989. "New AB Polyesters and a Polymethacrylate Containing Dipolar p-Phenyleneazo Groups," *Macromolecules*, 22:3525.

26. Everaerts, A., S. A. Roberts and H. K. Hall, Jr. 1986. "Synthesis and Properties of Semiconducting Aromatic Polyquinone Diimines," *J. Polym. Sci. Polym. Chem. Ed.*, 24:1703.

EPILOGUE

Many functional groups have been incorporated into macromolecular frameworks to design materials which will perform specific chemical functions. The treatment of different classes of functional polymers having differing end-use applications on the basis of constituent functional groups unifies apparently unrelated materials into a single chemical grouping. As evident in the present volume, many structural similarities which are not so obvious when they are discussed as thermally labile polymers, photochromic polymers and biodegradable polymers, are evident when the polymer constitution is viewed structurally on the basis of the characteristic groups.

In view of the rapid growth in various functional polymers, it is necessary to evaluate polymer constitution in terms of functional groups manifesting those properties. The azo functionality is an interesting group exhibiting a range of properties. Viewed in perspective, interest in azo functional polymers has been sporadic and uneven. As discussed in Chapter 2, the azo functionality exhibits a number of interesting properties. Much of the aromatic azo chemistry orginates from azo dyes, likewise much of the aliphatic azo chemistry from thermolabile azo initiators. The same scenario is extended into the polymer science as well. Polymers based on aliphatic moieties were studied purely for their thermally labile property, mostly in industrial laboratories during the sixties. No reports of photolysis, biodegradation, and other properties are known for the aliphatic azo functional polymers. In the case of aromatic azo polymers, work started on azo-based thermally stable polymers in the sixties, on photoresponsive polymers in the seventies and on biodegradable polymers only in the mid-eighties, and in most cases polymers have been designed using azobenzene nucleus only. Although photoisomerization behavior of other azoaromatic units such as azopyridines and azonaphthalenes is known, no studies are reported thus far using polymers containing azo-heterocyclic units.

Reactions such as o-metallation of azobenzene moieties constituting polymeric chains is an area worth exploring. The o-metallated complexes of backbone azo polymers may lead to some interesting organometallic polymers. The memory storage devices of the type described recently by Japanese workers (Chapter 2) are likely to lead to polymeric equivalents.

The field of drug delivery vehicles based on azoaromatic linkages is still very much in its infancy. A number of physicochemical, microbiological, biochemical and pharmacological aspects of drug delivery to the colon still remain to be established. Likewise, the optimal physical nature of the carriers also has to be explored, such as microspheres, microcapsules, nanoparticles, matrix devices, etc. The efficacy of peptide drug delivery remains very low and likewise the knowledge of peptide absorption through the GI tract is still fragmentary.

The investigations on the liquid crystalline azopolymers have begun actively only in the last decade and are likely to continue into the next decade. The field of polymeric dyes as food additives, it seems, has lost some momentum due to toxicity problems associated with the degradation products.

Although numerous patents and applications describe the possible industrial uses for azo-containing polymers, as far as the authors are aware, no industrial process exists. Polymeric azo dyes were marketed by Dynapol corporation but later discontinued. The literature does not seem to contain any studies on the health risks associated with handling of azo-polymers but some aliphatic azo polymers are reported to be explosive.

The search for such materials continues. Often this diligent, ongoing effort is met with a seemingly endless series of failures, but then there are occasional successes and the exhilarating steps forward—thus the pursuit continues.